T0205909

JOINT COGNITIVE SYSTEMS

Patterns in Cognitive Systems Engineering

JOINT COGNITIVE SYSTEMS

Patterns in Cognitive Systems Engineering

David D. Woods
Erik Hollnagel

CRC Press
Taylor & Francis Group
Boca Raton London New York

CRC Press is an imprint of the
Taylor & Francis Group, an **informa** business

CRC Press
Taylor & Francis Group
6000 Broken Sound Parkway NW, Suite 300
Boca Raton, FL 33487-2742

First issued in paperback 2019

© 2006 by Taylor & Francis Group, LLC
CRC Press is an imprint of Taylor & Francis Group, an Informa business

No claim to original U.S. Government works

ISBN-13: 978-0-8493-3933-2 (hbk)
ISBN-13: 978-0-367-86415-6 (pbk)

Library of Congress Cataloging-in-Publication Data

Catalog record is available from the Library of Congress

**Visit the Taylor & Francis Web site at
http://www.taylorandfrancis.com**

**and the CRC Press Web site at
http://www.crcpress.com**

Contents

Preface

The gadget-minded people often have the illusion that a highly automatized world will make smaller claims on human ingenuity than does the present one ... This is palpably false.

Norbert Wiener, 1964, p. 63

For almost 25 years, Cognitive Systems Engineering (CSE) has searched out and listened to stories of *claims on human ingenuity* as fields of practice have changed and adapted to new pressures, new devices, and new opportunities. In these stories, change challenged how activities are coordinated, how systems are resilient at boundaries, and how artifacts provide affordances.

This meant we could complement the previous book on joint cognitive systems, which focused on the foundational concepts, with a book that used stories of cycles of complexity and coping to show the main values, concepts and approaches of CSE in action. This book provides a way to look at our collective progress. The general storylines make up a base of findings and patterns that can be used to launch new studies of work, to project the effects of new rounds of change, to search out promising directions for innovating support. Ironically, thinking about representing a research base as a set of abstract narratives is in itself an exploration of how to support and enhance sharing technical, scientific and design information.

The stories told here are about more than the sharp end of practice. In parallel, each story tells another tale—revealing gaps and deficiencies in stories we tell ourselves about *our* relationship with technology, about how we accomplish *our* goals, and how *our* systems are vulnerable to breakdown. Acting in roles as researchers or designers or managers does not grant us any immunity from the processes and difficulties of coping with complexities.

One theme running through this volume is—*surprise*—that action is oriented to the future, anticipating or being prepared for what comes next. This book is intended to provide a start on a pattern base to help us be prepared to be surprised:

• As we observe joint systems at work in cycles of change and adaptation to come,

• As we envision with others how joint systems will work given new pressures and opportunities, and

• As we reflect on and revise our beliefs about work as one of the set of stakeholders whose beliefs support or hobble joint systems as they work.

David D. Woods   Erik Hollnagel

Chapter 1

Core Activities and Values

Four guiding values—authenticity, abstraction, discovery, participation—infuse the activities of Cognitive Systems Engineering. When these values are put into practice CSE discovers patterns in joint cognitive systems (JCSs) at work, abstracts patterns in how joint cognitive systems work, and, ultimately, designs joint cognitive systems that work.

ADAPTABILITY VERSUS LIMITS

Look at the news as organizations struggle to meet new demands on their performance, how to be more efficient under resource pressures, or how to recognize and exploit opportunities created as the technology baseline advances. Back and forth the comments cycle: Are people the solution or are people the problem to be solved? Is human performance the key to achieving goals or is human performance the limit to be overcome?

People: sinners or saints? An old debate to be sure, but a false position that continues to fail to improve performance or safety as it assumes the wrong unit of analysis and design. The path to success begins with recognizing the paradox that, simultaneously, we are both the source of success and of failure as we create, operate, and modify human systems for human purposes.

People—as adaptive agents, learning agents, collaborative agents, responsible agents, tool creating/wielding agents—create success under resource and performance pressure at all levels of the socio-technical system by learning and adapting to information about the situations faced and the multiple goals to be achieved (Hollnagel, 1998; Woods & Cook, 2002). This is one of the laws that govern joint cognitive systems (JCSs) at Work:

> Success in complex systems is created through work accomplished by joint cognitive systems (JCSs), not a given which can be degraded through human limits and error.

The usual rationalization for focusing on technology as a remedy for human performance problems is that people are limited. This misrepresents the situation, since all real cognitive systems at work are finite as Simon (1969) first emphasized.

To bring the full weight of this insight to bear, one must always keep the Bounded Rationality Syllogism in mind:

All cognitive systems are finite (people, machines, or combinations). All finite cognitive systems in uncertain changing situations are fallible. Therefore, machine cognitive systems (and joint systems across people and machines) are fallible.

The question, then, is not fallibility or finite resources of systems, but rather the development of strategies that handle the fundamental tradeoffs produced by the need to act in a finite, dynamic, conflicted, and uncertain world.

The core ideas of Cognitive Systems Engineering (CSE) shift the question from overcoming limits to supporting adaptability and control (Hollnagel & Woods, 1983; Woods & Roth, 1988; Woods & Tinapple, 1999; Hollnagel, 2001; Chapter 2 in Hollnagel & Woods, 2005).[1] The base unit of analysis is the Joint Cognitive System (JCS), not people versus technology; and the key process to study, model and support is how people cope with complexity. These core ideas have provided the basis for new findings on how JCSs work and new design concepts to support the functions of JCSs. This book builds on the previous volume on joint cognitive systems which articulates the foundations of CSE (Hollnagel & Woods, 2005). Here, we will explore some of the results of 25 years of CSE in action in terms of:

(1) Techniques to discover how JCSs work;
(2) Laws or first principles generalized from studies of JCSs at work;
(3) General demands on work that must be met or carried out by any JCS;
(4) Generic requirements or forms of support for the work of JCSs.

Work in CSE culminates in reusable design seeds that illustrate how the general requirements can be met through the design of JCSs that work (Woods, 1998).

Throughout the book are stories of particular work settings where people, technology and work intersect. The value of these stories lies in how they capture general recurring patterns about the resilience or brittleness of strategies in the face of demanding situations, about coordination or miscoordination across agents using their knowledge to pursue their goals, and about the affordances or clumsiness of artifacts in use.

COMPLEMENTARITY

It is, ... the fundamental principle of cognition that the universal can be perceived only in the particular, while the particular can be thought of only in reference to the universal.

E. Cassirer, 1953, p. 86

---

[1] In this book, cross-references are made repeatedly to the companion book *Joint Cognitive Systems: Foundations of Cognitive Systems Engineering*, Hollnagel and Woods, 2005, as *JCS-Foundations*, with reference to the chapter or pages that discuss the foundational concepts.

Fields of practice have a dual status in CSE. They are substantial specific settings that demand assistance to meet performance demands and resource pressures. Studying them and designing for them from this point of view seems "applied." Yet at the same time, these fields of practice function as "natural laboratories" where the phenomena of interest play out.

These two statuses are intimately coupled. From one perspective—research, "there are powerful regularities to be described at a level of analysis that transcends the details of the specific domain. It is not possible to discover these regularities without understanding the details of the domain, but the regularities are not about the domain specific details, they are about the nature of human cognition in human activity" (Hutchins, personal communication, 1992). A design perspective reverses the coupling, "if we are to enhance the performance of operational systems, we need conceptual looking glasses that enable us to see past the unending variety of technology and particular domains" (Woods & Sarter, 1993).

Coordinating these two strands defines complementarity (Figure 1). In one strand, Discovering Patterns in Cognition at Work, inquiry is directed at capturing phenomena, abstracting patterns and discovering the forces that produce those phenomena despite the surface variability of different technology and different settings. In this sense, effective research develops a book of "patterns" as a generic but relevant research base.

But the challenge of stimulating innovation goes further. A second strand of processes is needed that links this tentative understanding to the process of discovering what would be useful. Success occurs when "reusable" (that is, tangible but relevant to multiple settings) design concepts and techniques are created to "seed" the systems development cycle (Woods & Christoffersen, 2002).

Discovery of what would be useful occurs in the research cycle because development also creates opportunities to learn. Artifacts are not just objects; they are hypotheses about the interplay of people, technology and work. In this cycle prototypes function as tools for discovery to probe the interaction of people, technology and work and to test the hypothesized, envisioned impact of technological change.

CSE in the JCS synthesis is intensely committed to building a theoretical base through investigations and design interventions into ongoing fields of practice, at the same time that it innovates new designs of artifacts, representations and cooperative structures. This is a cycle of empirical investigation, modeling, and design, where each activity stimulates the other in a complementarity between the particular and the universal. Fields of practice are applications, but also function as natural laboratories. Technological change is a tool to derive general lessons about cognition and collaboration, while these lessons about the dynamics of cognitive systems act as a guide to help us see the essential factors behind the wide variability of work and technology in particular contexts.

The tension between the universal and the particular as expressed in the epigraph of this section is basic to any approach to cognition at work. As a result, CSE cuts across normal categories and links together—technological and behavioral sciences, individual and social perspectives, the laboratory and the field, design activity and empirical investigation, theory and application, in a new complementarity at the intersection of people, technology and work.

CORE VALUES OF CSE IN PRACTICE

One way to understand a field of practice, be it operational, engineering, or scientific, is to identify its core values. While usually implicit in a community and often violated, they provide an inside-out view of the culture. The core values of CSE emerge from the base activities needed to study and design JCSs (Figure 1).

Observation. CSE is committed to observing work in context.
CSE starts with observation of worlds of work in many different ways. In observation we work with practitioners to understand how they adapt to the pressures and demands of their field of activity.

Our methods and our debates about our methods are grounded in the need to observe and understand work in context (Lipshitz, 2000). The key referent is always where and how did you observe work? Methods vary as different ways to shape conditions of observation with different uncertainties about what was noted (as observation is never neutral). Observation in context generates debates about the nature of work—what are the cognitive demands, how is work coordinated across multiple parties and roles, how do broader levels (organizational context) influence work at the level of interest. Observation of work in context would seem to mean that one only looks at the "sharp" end of practice—the roles and activities which make contact with the process to be monitored and controlled (Figure 2). But all work—management, research, design—can be seen as a field of practice, a sharp end in itself that exists in its own organizational contexts that place performance pressures on that activity.

One value for CSE is authenticity—how are your claims about JCSs at work based on or connected to authentic samples of what it means to practice in that field of activity and how the organizational dynamics pressure or support practice. The search for authenticity stimulates converging processes of observation. Authenticity refers to the forms and limits on access to practice, how one shapes the conditions of observation, and how one understands the processes of change ongoing in worlds of practice. There are many labels for this core activity; some refer to it as situated cognition and others as a call to adopt ethnographic methods.

To state this activity succinctly as a core value: CSE is concerned with how to transcend limits to authenticity in order to capture how the strategies and behavior of people are adapted to the constraints and demands of fields of practice.

Abstraction. CSE is committed to abstracting patterns across specific settings and situations.
The risk of observation in context is that the observer quickly can become lost in the detail of particular settings at particular points in time with particular technological objects. It is by comparison and contrast across settings over time that one can abstract patterns and produce candidate explanations for the basic

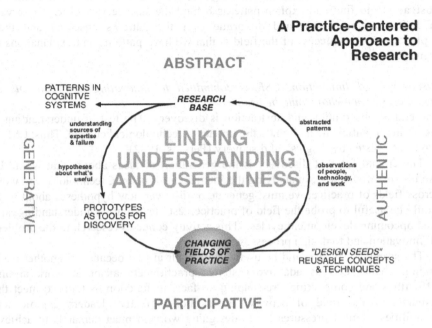

Figure 1: Complementarity defines Cognitive Systems Engineering: A setting is simultaneously a significant field of practice undergoing change and a natural laboratory where general patterns play out. Note the four basic activities around the cycle: observation, abstracting patterns, understanding expertise, promising hypotheses about what would be useful. These make up two processes around the cycle: a top half of understanding patterns in JCSs and a back half of innovating promising design seeds by using prototypes as tools for discovery. Each of the four basic activities should express two of the four basic values marked around the outside of the cycle.

patterns. Through abstraction we build up a set of the phenomena on JCSs at work as generalizations and as patterns in need of explanation. Plus, the pattern base of generalized phenomena becomes a means to jump start observations in a new particular context. One main goal of this book is provide a compact discussion of some of the patterns, concepts, and phenomena of JCSs at work—to look at what we think CSE has learned to this point.

In the study of JCSs at work, we are trying to understand how artifacts shape cognition and collaboration given the organizational context and problem demands in fields of practice, and how practitioners, individually and as groups, informally and formally, shape artifacts to meet the demands of fields of activity within the pressures and resources provided by larger organizations (Figure 3).

To state this activity succinctly as a core value: CSE confronts the **challenge of abstraction** to find and explain patterns behind the surface variability. Observers of JCSs in context can and do argue over the patterns observed and their explanations—the success of the field is that we have patterns and explanations to argue about.

Discovery and Innovation. CSE is committed to innovating new concepts as hypotheses about what would be useful.
The end of observation and abstraction is discovery. The test of understanding is the ability to anticipate and guide the impact of technological change. Thus, CSE is driven and tested by "logics" of discovery (Perkins, 1992).

The crucial form of discovery is the generation of ideas about what would be useful—*promisingness*. In studying the interaction of people, technology and work across fields of practice, we must generate or discover new hypotheses about what would be useful to probe the field of practice, test our tentative understanding, and seed upcoming development cycles. This activity connects research to the problem of innovation and to design processes.

The search for what could be useful is difficult in part because of another basic finding about JCSs as adaptive systems: practitioners adapt to work around difficulties and complexities, re-shaping artifacts to function as tools to meet the demands of the field of activity. In addition, effective leaders exploit new capabilities to meet pressures for re-designing work to meet demands to achieve more and achieve it more efficiently (see Law of Stretched Systems in chapter 2).

Nevertheless, the point of the processes of observation, abstraction and explanation is to find the essential factors under the surface variability. And the strongest test of how well a process is understood is its ability to anticipate and guide the multiple impacts of technological change on JCSs at work.

To state this activity succinctly as a core value: CSE is successful to the degree that it **sparks inventiveness** to discover new ways of using technological possibilities to enhance work, to identify leverage points, to seed development with promising directions, and to minimize negative side effects of change.

Participatory. CSE participates in the struggle of envisioning with other stakeholders to create the future of work.
In the final analysis, CSE activities connect back to processes of change in ongoing fields of practice (Woods & Dekker, 2000). This leads to a dramatic risk for the researcher—they are part of the process under study as their work and results (or others' interpretation of their results) is part of the change process. The researcher is not neutral, distant or detached. Rather, the researcher is a participant in the struggle to envision and create future operational worlds with other stakeholders. Researchers also must acknowledge their role as designers—the development of tools "that make us smart or dumb" (Norman, 1993), and their responsibility for failures as new artifacts generate surprising reverberations and adaptations (Cook & O'Connor, 2005).

Because prototypes-as-objects embody hypotheses about what would be useful, the designer functions as experimenter. The possibilities of technology afford designers great degrees of freedom. The possibilities seem less constrained by questions of feasibility and more by concepts about how to use the possibilities skillfully to meet operational and other goals. The adaptive response of people and organizations to systems tests the hypotheses about what would be useful embodied by particular prototypes or systems (Woods, 1998). As a result, the designer needs to adopt the attitude of an experimenter trying to understand and model the interactions of task demands, artifacts, cognition, collaboration across agents, and organizational context.

An experimental stance in design means:

• Design concepts represent hypotheses or beliefs about the relationship between technology and cognition/collaboration,
• These beliefs are subject to empirical jeopardy by a search for disconfirming and confirming evidence,
• These beliefs about what would be useful are tentative and open to revision as we learn more about the mutual shaping that goes on between artifacts and actors in a field of practice.

This stance is needed, not because designers should mimic traditional research roles, but because this stance makes a difference in developing systems that are useful—avoiding the result of building the wrong system right (Woods, 1998).

To state this activity succinctly as a core value: CSE helps **create future possibilities** as participants with other stakeholders in that field of practice.

ON SYSTEMS IN CSE

CSE is a form of systems engineering (Hollnagel & Woods, 1983). Taking a systems perspective has three basic premises.

1. *Interactions and emergence*: a system's behavior arises from the relationships and interactions across the parts, and not from individual parts in isolation.
2. *Cross-scale interactions* (multiple levels of analysis): understanding a system at a particular scale depends on influences from states and dynamics at scales above and below.
3. *Perspective*: how the parts of a system and levels of analysis are defined is a matter of perspective and purpose (*JCS-Foundations*, p. 115-116).

The unit of interest and analysis in CSE, as one perspective on complex systems, is those factors, processes, and relationships that emerge at the intersections of people, technology and work. These emergent processes cannot be seen if one only looks at any one of these alone. Traditional disciplines that are related in part to cognition at work separate the people, the technology, and the work setting into their own units of analysis as the outsider takes a psychological, technology, or domain perspective alone. In CSE the interactions across these three are the phenomena of critical interest. The relationship is one of mutual adaptation where people as goal directed agents adapt given the demands of work settings and the affordances of the artifacts available (Woods, 1988). This is a case of agent-environment mutuality, after von Uexkull (1934) and Gibson (1979), which means that each can be understood only in relation to the other and that each changes through adaptation to the other (cross-adaptation). Characteristics of artifacts and tasks interact to place *demands* on the activities of and coordination across agents. Characteristics of tasks and needs for coordination across agents interact to specify how artifacts *support* (or fail to support) work. Properties of artifacts and agents interact to generate how *strategies* are adapted for work in the task domain.

What emerges at the intersections of people, technology and work are:

• Patterns in **coordinated activity**—or its contrast, miscoordination: how cognitive work is distributed and synchronized over multiple agents and artifacts in pace with changing situations.
• Patterns in **resilience**—or its contrast, brittleness: the ability to anticipate and adapt to potential for surprise and error.
• Patterns in **affordance**—or its contrast, clumsiness: how artifacts support (or hobble) people's natural ability to express forms of expertise in the face of the demands on work.

As a kind of systems perspective, CSE is also concerned with cross-scale interactions—and one can refer to these interactions as an interplay between the 'sharp' and 'blunt' ends of a field of practice as in Figures 2 and 3 (Reason, 1990; Woods et al., 1994; Cook et al., 1998). At the sharp end of a complex system, practitioners, such as pilots, spacecraft controllers, and, in medicine, nurses, physicians, technicians, pharmacists, directly interact with the hazardous process. Groups of practitioners are successful in making the system work productively and safely as they pursue goals and match procedures to situations. However, they do much more than blindly follow rules. They also resolve conflicts, anticipate hazards, accommodate variation and change, cope with surprise, work around obstacles, close gaps between plans and real situations, detect and recover from miscommunications and mis-assessments (Figure 3).

At the blunt end of the system, regulators, administrators, economic policy makers, and technology suppliers control the resources, constraints, and multiple incentives and demands that sharp end practitioners must integrate and balance to accomplish goals. Downward, coordination, resilience, and affordances are affected by how organizational context creates or facilitates resolution of pressures/goal conflicts/dilemmas. For example, by mismanaging goal conflicts or introducing clumsy automation the blunt end can create authority-responsibility double binds for sharp end practitioners (Woods et al., 1994; Woods, 2005a). Upward, coordination and resilience are affected by how the adaptations of local actors in the form of workarounds or innovative tactics reverberate and influence more strategic goals and interactions (e.g., workload bottlenecks at the operational scale can lead to practitioner workarounds that make management's attempts to command compliance with broad standards unworkable; see Cook et al., 2000; Woods & Shattuck, 2000).

Stories of how sharp end practice adapts to cope with the complexities of the processes they monitor, manage and control are incomplete without accompanying stories of cross-scale interactions about how the resources and constraints provided by the blunt end of the system support or undermine sharp end practice (Cook et al., 1998). In the other direction, work always occurs in the context of multiple parties and interests as moments of private cognition punctuate flows of interaction and coordination (Hutchins, 1995a). From the perspective of cross-scale interactions, then, one can think of the CSE strategy of focusing on how people in various roles cope with complexity as a "middle-out" analysis connecting processes of coordination, resilience, and affordance with blunt end influences and with aspects of private cognitive processes that occur within individual heads (see the discussion of the third family of laws in Chapter 12).

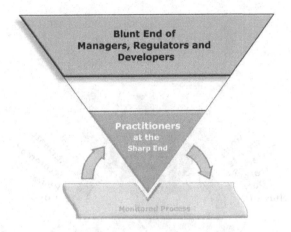

Figure 2. The sharp and blunt ends of a complex system. Understanding how practitioners' adapt also requires understanding how the organizational context supports or adds to the goal conflicts, dilemmas and double binds experienced at the sharp end (adapted from Woods et al., 1994).

Studies of JCS at work capture relationships across demands, affordances, and strategies. These interact in dynamic and cross-adaptive processes to express stories of JCSs at work: stories of coordination or miscoordination, stories of resilience or brittleness, stories of support or clumsiness from artifacts.

The stories that capture these relationships of interest in JCSs are all expressions of the dynamic interplay captured by Neisser's (1976) perceptual cycle (see Figure 1.7 in *JCS-Foundations*, p. 20 for the central role of this concept in understanding JCSs at work). In fact, most of the diagrams used in this book are cyclic, and are variations on the fundamental dynamic captured in Neisser's cycle. Furthermore, these cycle diagrams often work at two levels, with the explicitly depicted cycle being built on a set of underlying cycles.

A good example of this occurs in Figure 1. The four core activities of CSE as captured in this diagram are not isolated steps carried out in series. What is productive comes from working the inter-linkages between core activities in mini-perceptual cycles around the main cycle. Otherwise, for example, observation becomes mere description and cataloging rather than part of the process of tentative abstraction. Or abstraction becomes mere naming and categorization—substituting general labels for capturing the dynamics across people, technology and work activities in JCSs. The focus of this book is on how the interplay across these mini-perceptual cycles in CSE research and design have identified general patterns about JCSs at work.

Figure 3. At the sharp end of a complex system: the interplay of problem demands and practitioners' coordinated activity govern the expression of expertise and failure. The resources available to meet problem demands are provided and constrained by the organizational context at the blunt end of the system (adapted from Woods et al., 1994).

It is very important to see the contrast between two perspectives: the CSE perspective, as a form of systems analysis that uncovers stories of resilience, coordination, and affordance in the JCS at work; and the technological perspective, which focuses on new or expanded capabilities as automation or algorithms are realized and deployed. The latter perspective, automation or algorithm based, generates stories about the possibilities for new levels of autonomy in machine agents, and presents work as if it consisted only of an agent and a task in isolation. The stories of work generated from taking this perspective describe how human roles support the new algorithm—how people need to follow the algorithm, accept the results from machines that implement algorithms, or provide data input and other services so that the algorithm can be run on real world cases (Woods, 1986a and b). The latter perspective also generates many realization-based stories about work in the future—stories that presume hypothesized benefits will follow (and only those effects will occur) if only sufficient investment is made to advance the technological baseline on one or another dimension (Woods & Dekker, 2000). These stories block out observation and analysis of how people will adapt to carry out roles as responsible agents in order to achieve the goals of practice.

The technology- or algorithm-centered perspective finds human behavior very erratic and perplexing as the sharp end of practice continues to spill over the prescribed borders envisioned for practitioners. But when we adopt the perspective of JCSs as adaptive control systems, we begin to recognize the regularities, the basic patterns, and even the "laws" at work. For example, one reason algorithm-based stories of future work prove so very wrong is that they violate the following basic principle or law (Woods, 2002):

> There is no such thing as a *cognitive vacuum*: When we [as designers] create, represent, or shape some experience for others, those people learn something [even if it is not what we intended]; form some model or explanation [even if it is not veridical]; extract and index some gist [even if it is quite different from what was relevant to us]; see some pattern [even if we encoded other relationships or no relationships at all]; pursue some goal [even if it is not the goal we most value]; balance tradeoffs [even if our idealizations about work push them out of our view].

When we assume that roles in future work systems will form around how to run and support the algorithms we deploy as automation, we will be surprised by the adaptive behavior of the JCS at work (Woods et al., 2004). When design does not provide the information and external representations to help convey a model of how the device or process works, the result is not lower workload or simpler interfaces for users as is often claimed. Since people fundamentally build explanations for the processes they confront everyday (Bruner, 1990), we can quite accurately predict what happens when design of artifacts does not help users build accurate models, at some level of analysis, of the relevant processes: users will develop many diverse models of the process or device and these models will be quite wrong (e.g., Cook, Potter, et al., 1991).

PATTERNS

Readers may notice the abundant usage of pattern-words in the introduction up to this point: the research base is a book of generic but relevant patterns; studies point to general demands in forms of work that must be met or carried out by any JCS; linking understanding to design generates generic requirements to support forms of work; innovation provides promising design "seeds" that can be reused in multiple development projects across different technology and settings.

The pattern-book approach to capture the results of CSE derives directly from Christopher Alexander's concept of patterns in design.

Each pattern describes a problem which occurs over and over again in our environment, and then describes the core of the solution to that problem, in such a way that you can use this solution a million times over, without ever doing it the same way twice. (Alexander et al., 1977, p. x)

Patterns are a set of relationships, and the relationships can recur even as the elements change. There are relationships across patterns, as patterns are cross-connected. Patterns are multi-level, as a pattern is also a sub-part in larger patterns and a context for lower level patterns.

These patterns are empirical generalizations abstracted from and grounded on observations made through different studies (see Figure 5). Patterns are a form of diagnosis that capture the essence of a problem (a hypothesis open to revision) and point to directions to be pursued to resolve the situation diagnosed (specify what could be promising directions to search for new solutions).

Patterns become guides to understand new specific situations, yet investigators can judge for themselves how each pattern covers the situation at hand or what revisions are necessary. Patterns are open-ended as they point to solution directions, yet allow participants to solve each problem for themselves by adapting the general seed or technique to the unique combination of conditions in the context at hand.

Note on terminology:
As was noted in *JCS-Foundations* (pp. 2-3), in CSE we do not describe users, clients or customers; rather we refer to *stakeholders* and *practitioners*. The term *problem-holder* emphasizes authority and responsibility—the person who holds a problem is one who has some scope of authority to resolve the situation and also who is responsible for the outcomes that follow. And, what one is responsible for influences what kinds of problems one recognizes. The term *practitioner* emphasizes the experience people build up in their roles and how their activities are adapted for achieving the multiple goals primary to those roles. Both terms emphasize the connection of human activity to goals and constraints in the world that is central to the character of JCSs.

To summarize: Core Activities and Values

CSE is a systems approach to phenomena that emerge at the intersection of people, technology and work. CSE draws on observations of JCSs at work; previous general patterns are brought to these observations; contact with work leads to tentative abstractions about what patterns are present; this understanding stimulates and guides the search for what would be useful to support the kinds of work seen; these insights stimulate innovation of design seeds in participation with others to support the work and processes of change.

CSE, as a practice-centered research and design process, is generic but relevant. It finds in the particular the existence and expression of universal patterns. It tests and revises these patterns by seeing design concepts as tentative hypotheses subject to empirical jeopardy. These basic activities and values are not contradictions or conflicts, but creative tensions at the root of complementarity, harnessed for innovation.

Additional resources related to these concepts can be found in the multi-media production, *Watching Human Factors Watch People at Work* (url: http://csel.eng.ohio-state.edu/hf99), and in treatments by Winograd and Flores (1986), Hutchins (1995a), Hollnagel (1998), and Woods & Christoffersen (2002).

Discovering Patterns in Joint Cognitive Systems at Work

The study of joint cognitive systems in context is a process of discovering how the behavior and strategies of practitioners are adapted to the various purposes and constraints of the field of activity. To study joint cognitive systems requires going behind the more or less visible activities of practitioners to uncover how these activities are parts of larger processes of collaboration and coordination. The key is to uncover how people adapt to exploit capabilities and workaround complexities as they pursue their goals.

A JCS AT WORK

In the year after the Three Mile Island accident, a team of what we would today call cognitive system engineers was observing how nuclear power control room operators handle simulated faults in a full scope training simulator (Figure 4). Over several simulated cases the observers noticed that one of the operators would move to the reactor display/control area, but they couldn't tell how he knew to check out information about these systems. The observers asked the operators about what triggered them to monitor these displays but the operators made only vague general comments, intensifying the mystery.

Eventually, the observers recognized that the operators were picking up changes in clicking sounds made by mechanical counters that indicated the position of the control rods as they moved in and out of the reactor. It turned out that activity of the automatic rod control system was indicated by the sounds made by the mechanical counters as the control system moved the rods to regulate reactor power and to shut down the reactor if faults occurred.

Following this recognition the observer team watched more closely in different situations where rod control was part of the case evolution. Operators didn't move to investigate any or all clicking sounds from the rod control system; rather they only moved to investigate a change in sounds that didn't fit their expectations. When the rod control system was active (clicking) and operator didn't expect much activity, he would begin to investigate. When the rod control system was inactive (not clicking) and operator did expect activity, he would begin to investigate. When changes in rod control system activity and the associated clicking sounds matched expectations, the operators went on about other tasks in the control room and there were no observable signs that the operators were monitoring or thinking about rod control.

When the observers tested their explanation by asking the operators if they listened for unexpected changes in the clicking sounds, the operators looked a little surprised and impatient at the observers. Of course, they then remarked, the sounds were good clues as to what the rod control system was doing and helped them know when something funny was going on in that part of the process.

Figure 4. A nuclear power plant control room during a simulated fault (training simulator). Note: data from each sensor is presented in a single display; the displays are spatially distributed, but some computer based displays have been added. In addition, note the annunciator or alarm lights at the top and the status panel below them. To the experienced operator, these indicate that the plant is in the middle of a major emergency—the plant should be shut down; emergency cooling should be running to compensate for several disturbances in several basic functions (From D. Woods, personal collection).

Chapter 2

Joint Cognitive Systems Adapt to Cope with Complexity

General patterns in cognitive work do not exist separate from particular intersections of specific people, technologies and work activities. Hence, field settings are more than unique domains or exotic fields of practice, but function in parallel as natural laboratories where these generic patterns emerge. What approaches would help us discover the essential patterns underlying the surface diversity of people, technology and work? What techniques would help us begin to unpack or partially decompose the complexities of systems of people, technologies and work activities into meaningful parts and their interactions? How do we use general patterns to guide change and innovate new possibilities in concert with other stakeholders? This chapter introduces the basics of how to shape the conditions of observation, the consequences of the Law of Fluency, and the central role of the Law of Demands in studying JCSs.

> Modern biology has demonstrated the enormous power of functional analysis coupled with naturalistic observation.
>
> U. Neisser, 1991, p. 35

ADAPTATION IN JOINT COGNITIVE SYSTEMS AT WORK

To discuss how to observe and discover, we must first prepare to be surprised. This preparation begins with existing abstractions about JCSs at work. This section covers generalizations from past work on how JCSs are adaptive systems.

Observational studies of JCSs in context have built a body of work that describes how technology and organizational change transforms work in systems. Points of technology change push cycles of transformation and adaptation (e.g., Carroll's task-artifact cycle; Carroll et al., 1991; Winograd and Flores, 1986; Flores, Graves, Hartfield, & Winograd, 1988).

The review of the impact of new technology in one operational world effectively summarizes the general pattern from many studies (Cordesman & Wagner, 1996, p. 25):

> Much of the equipment deployed...was designed to ease the burden on the operator, reduce fatigue, and simplify the tasks involved in operations. Instead, these advances were used to demand more from the operator. Almost without exception, technology did not meet the goal of unencumbering the personnel operating the equipment... Systems often required exceptional human expertise, commitment, and endurance...There is a natural synergy between tactics, technology, and human factors...Effective leaders will exploit every new advance to the limit. As a result, virtually every advance in ergonomics was exploited to ask personnel to do more, do it faster and do it in more complex ways...One very real lesson is that new tactics and technology simply result in altering the pattern of human stress to achieve a new intensity and tempo of operations. [Edited to rephrase domain referents generically]

This statement could have come from studies of the impact of technological and organizational change in health care or air traffic management or many other areas undergoing change today (see Billings, 1997, and Sarter & Amalberti, 2000, for the case of cockpit automation). The pattern illustrates a more general law of adaptive systems that has been noted by several researchers (e.g., Rasmussen, 1986; Hirschhorn, 1997).

> *The Law of Stretched Systems:* every system is stretched to operate at its capacity; as soon as there is some improvement, for example, in the form of new technology, it will be exploited to achieve a new intensity and tempo of activity.

Under pressure from performance and efficiency demands, advances are consumed to ask operational personnel "to do more, do it faster or do it in more complex ways" (see NASA's Mars Climate Orbiter Mishap Investigation Board report, 2000, for a example).

Overall, the studies show that when new "black box" technology (and accompanying organizational change) hits an ongoing field of practice the pattern of reverberations includes (Woods & Dekker, 2000; *JCS-Foundations*, pp. 103-105):

• New capabilities, which increase demands and create new complexities such as increased coupling across parts of the system and higher tempo of operations,
• New complexities when technological possibilities are used clumsily,
• Adaptations by practitioners to exploit capabilities or workaround complexities because they are responsible to meet operational goals,
• Complexities and adaptations that are surprising, unintended side effects of the design intent,

- Failures that occasionally break through these adaptations because of the inherent demands or because the adaptations are incomplete, poor, or brittle,
- Adaptations by practitioners that hide the complexities from designers and from after-the-fact reviewers who judge failures to be due to human error.

The above patterns point to general laws of adaptive systems. For example, the foundation is Ashby's original Law of Requisite Variety (*JCS-Foundations*, pp. 40-47). In its most compact form: Only variety can destroy variety.[2] This quickly leads to a definition of skill or expertise as the ability to adapt behavior in changing circumstances in the pursuit of goals—context-conditioned variability.[3]

Discovery of patterns in JCSs at work begins with studying what the system is or has adapted to. The nuclear control room story on pages 15-16 illustrates how the demands for shifting attention in an event-driven, multi-threaded situation were related to a strategy of experienced operators—how they had learned to utilize an incidental consequence of the choice of mechanical counters for display of rod position. Serendipitously, the control room designers had provided an auditory display that served as an affordance to help operators meet the demands for shifting attention as situations changed (Woods, 1995b). Changing the specific display mechanism or interface would remove this valuable cue, yet the artifact seen as affordance for skilled control of attention provides a design seed that can be accomplished in many ways to meet the general demand of work.

Years later, in the context of problems with mode transitions in highly automated cockpits (e.g., Sarter & Woods, 1995), Nadine Sarter deliberately designed cues to serve this directed attention function for human-automation teamwork in several different modalities (peripheral visual, auditory and tactile) and found that tactile cues were particularly effective means to support directed attention skills (Sklar & Sarter, 1999; Nikolic & Sarter, 2001; Sarter, 2002; Ho et al., 2004). She developed in a principled and empirical process the same affordance that was illustrated, almost by accident, in the nuclear control room observation noted at the beginning of this section.

This case leads us to two laws of JCSs as adaptive systems critical for studying JCSs at work. The first is:

Law of Demands: What makes work difficult, to a first approximation, is likely to make work hard for any JCS regardless of the composition of human and/or machine agents.

---

[2] Ashby (1957 p. 200) used this example: "If it [the automatic pilot] is a good regulator the passengers will have a smooth flight whatever the gustiness outside. They will, in short, be prevented from knowing whether it is gusty outside. Thus a good pilot acts as a barrier against the transmission of information." Hence, the variation in the pilot's behavior can be said to 'destroy' the variation in wind speed.

[3] This is taken from work in perceptual-motor skills and research in ecological perception after Gibson, 1979. It follows directly from the Law of Requisite Variety. Ironically, matching variability in control to variability in the world produces stability, though underneath stability is a sea of variation, coordinated and adapted.

This law captures an essential attribute of CSE. CSE begins by observations and abstractions that identify what makes work difficult for any agent or set of agents; these define kinds of general demands in forms of work that must be met or carried out by any JCS. CSE is not trying to model processes in the head (Hollnagel, 2001; 2002); instead it is trying to model constraints and demands in the worlds of work given that cognitive agents using various artifacts will attempt to cope with those demands as they pursue their goals.

To take one kind of demand factor, situations vary in the demands they place on reorienting attention to new potentially relevant stimuli (Woods, 1995b). It is hard to manage attentional focus in changing situations (the demand for control of attention). One can over-focus, being too insensitive to new information or signals, or one can shift attention too easily, losing any coherence in responses as the situation evolves (also called vagabonding). Recognizing that this general difficulty is manifest in a particular situation is a key to finding leverage points for what would be useful and to discovering how to test for promising or fruitful design seeds (e.g., Sarter's studies of tactile cues in situations with high demands on control of attention).

Also notice that starting with general demands or difficulties moves analysis and design away from any contrast between what computers do and what people do (the false trail of function allocation; Hollnagel, 1999; *JCS-Foundations*, pp. 121-125). What makes work difficult confronts any and all systems of agents. The productive question is how to design the JCS to meet the demands. Substituting a machine role for a human role in itself simply fails to address this and, as a result, usually leads to new complexities and surprising adaptive behaviors in JCSs.

Identifying what makes work difficult will help us see how agents have or will adapt and how artifacts have affordances. This leads us to another law fundamental to JCSs at work:

Law of Fluency: "Well"-adapted work occurs with a facility that belies the difficulty of the demands resolved and the dilemmas balanced.

Through adaptation, JCSs hide the constraints adapted to. This is the general case for effective control as Ashby noted in his basic Law of Requisite Variety (*JCS-Foundations*, pp. 40-41). The worlds of work come to us as an adaptive web consisting of a complex conglomerate of interdependent and dynamic processes including complex artifacts, dynamic worlds, cognitive activity, coordination and conflict, and organizational dynamics. These pre-existing adaptive webs are being stretched and changed as performance demands increase, as resource pressure intensifies, as organizations reconfigure, and as the power of computation and connectivity expand.

This means that methods for studying JCSs uncover the demands and dilemmas of work that guided previous adaptation and which, when identified, can help anticipate how future processes of mutual adaptation will unfold as organizations and artifacts change.

The above tells us that studying work in context is a process of functional analysis and synthesis—capturing how the behavior and strategies of practitioners are adapted to the various purposes and constraints of the field of activity. In studying a workplace we are out to learn how the more or less visible activities of practitioners are parts of larger processes of collaboration and coordination, how they are shaped by the artifacts and in turn shape how those artifacts function in the workplace, and how their activities are adapted to the multiple goals and constraints of the organizational context and the work domain (Rasmussen et al., 1994; *JCS-Foundations*, pp. 129-134).

How then can we advance our understanding of only partially decomposable, mutually adapted, and dynamic complexes of factors? Understanding what an environment affords to agents (given their goals), how agents' behavior is adapted to the characteristics of the environment, and how this linkage changes is functional synthesis. Hence, the study of cognitive work in context develops a functional model that captures how the behavior and strategies of practitioners are adapted to the various purposes and constraints of the field of activity.

Functional synthesis is a process that coordinates multiple techniques in order to unpack complex wholes to find the structure and function of the parts within the whole. A functional synthesis analyses and breaks apart adaptive processes in order to develop a model that explains the structure in relation to the purposes, where a function specifies how structures are adapted to purposes. Thus a functional model serves as an archetypical pattern and narrative generator that guides how specific stories can play out in multiple situations and settings.

To understand the issues about discovering how a JCS works, we need to consider a specific field of practice and examine a functional synthesis that captures some of the demands of that work setting. Doing this requires considering many details selected, and organizing them so that others can see how the JCS in question is adapted, and to what. The specific case is a study of the Intensive Care Unit developed by Richard Cook (1998) and called "Being *Bumpable*." The synthesis and the details will prove to be valuable resources to refer back to both for the discussion of method here and for later discussions of general patterns in JCSs at work.

To summarize: *Adaptations Directed to Cope with Complexity*

Studying and modeling a JCS at work requires understanding how the system adapts and to what it is adapted. The three guiding generalizations are the Law of Stretched Systems, the Law of Fluency, and the Law of Demands.

Chapter 3

Being *Bumpable*:
Consequences of Resource Saturation and Near-Saturation for Cognitive Demands on ICU Practitioners

by Richard I. Cook, University of Chicago

A case of synthesis of cognitive system functions—Richard Cook's
(1998) **Being "Bumpable"** *study of Intensive Care Units—illustrates*
the techniques and difficulties as one sets out to discover how real
JCSs function. This is a case that unlocked aspects of the nature of
practice. The starting point is to model the significance of insiders'
use of a highly coded term of practice. Cook uses this term and a
critical incident—a breakdown in the normal adaptive processes—as
wedges to break through the fluency of actual operations, uncovering
the demands on work and revealing how the system has adapted to
cope with these complexities. Understanding what it means to be
"bumpable" *and what the role of the* **bedmeister** *is, allows Cook to*
reveal the adaptive power of the intensive care unit as a JCS.

THE STORY: A DELAY

The case, part of studying hospital intensive care units, appears very simple on the surface.

A patient undergoing surgery needed to go to an intensive care unit when the surgical procedure was complete. The surgical procedure was finished before the intensive care unit bed was available, so the patient was taken first to the recovery room and transferred to the intensive care unit after a few hours.

This description is of course incomplete. Over these few hours there were telephone calls and discussions, some disagreements and disputes, extra work, and changes of plan. These sorts of events are common in most hospitals: something

23

takes less or more time than expected and the organization responds by diverting the flow of activity in such a way as to handle the early or late arrival of a patient.

Several features of the episode are noteworthy. The fact that a patient has surgery without having the resources available to take care of him after surgery seems unusual. We do not think of health care as being so haphazardly organized. It is more striking that this sort of thing is common. Practitioners will tell you that coping with such situations is routine and unremarkable. This sort of event is, in large part, business as usual in health care, and scarcely deserves notice.

The simple story of this episode reveals little about why these situations occur or how they are recognized and handled by the workers. Nothing is included in the story about the organization that handles events or what such events may mean in organizational terms. What are their possible consequences? What does it mean when the normal organizational processes break down in this way?

Here the individual workers have stepped in to make up for the limitations of the larger organization. Is this a kind of breakdown? The workers will tell you that this sort of situation occurs so often that we would necessarily conclude that the system is almost always broken.

Who is responsible for this state of affairs? Is it the workers themselves? What kinds of tradeoffs are made in the process of handling such an event? If this is as ordinary as it appears, why isn't it routinely prevented? Are there risks associated with these activities? If so, to whom does the risk accrue? Who benefits? Who pays? Do things always "work out" for the best or are these events somehow connected to accidents? Do the same sorts of things happen elsewhere in the hospital? To answer these sorts of questions requires that we look more closely at the technical work that takes place in the ICU and the hospital and, in particular, at the ways that this work acts as a kind of organizational "glue" to keep the parts of the system working together.

THE INTENSIVE CARE UNIT
THE SCENE, THE CAST, AND BACKDROP

An intensive care unit (ICU) is a place for patients whose precarious condition requires moment-to-moment attention by nurses, and by advanced technology, particularly mechanical ventilators. This sort of care requires close, often invasive (i.e., skin penetrating) monitoring equipment and specialized worker knowledge.

Within the hospital, patients typically come to the ICU from the operating room after surgical procedures that can be expected to leave them in precarious circumstances—conditions frequently described as "unstable"—or from the emergency room or, sometimes, from the ward (sometimes called "the floor") after deterioration of their condition (e.g., after cardiac arrest). Large hospitals may have multiple ICUs with different characteristics corresponding to specialty: cardiac surgical, neurosurgical, medical, burn, and pediatric ICUs are common. The differentiation between these ICUs reflects the increased complexity of modern medical care.

An ICU is an expensive place to receive care, usually an order of magnitude more costly than care on the floor. This reflects the costs of the one-to-one nurse-to-patient ratio of an ICU but also the presence of specialized equipment that is needed for the care of such sick patients. ICUs vary in their sophistication across hospitals, but within any facility the local ICU is the site of the most advanced care possible for that facility. It is not surprising, therefore, that the ICU in a busy hospital is an important resource and is typically a place where ratio of nurses to patients is closest to 1:1.

The floor and the ICU represent extremes in sophistication and intensity of care. Some hospitals also have intermediate level of care units, often called "stepdown" units, i.e., places where the sophistication of care is a step down from that found in the ICU but greater than that found "on the floor." The nursing ratio is lower here than in the ICU, e.g., 1:4 or 1:5, but still greater than that found on the floor, where the ratio may be 1:15 or even lower. The intermediate level of care found in a stepdown unit makes it possible to accommodate a patient who is too sick to be cared for on the floor but not so ill as to require care in the ICU.

The variety of level-of-care units provides a variety of options for the care of individual patients, but it also creates a burden. Someone must evaluate each new patient and sort out who should go where. Someone must re-evaluate patients at intervals to determine if their condition has improved (or worsened) so that they should receive a different level of care. The evaluation process is seldom a matter of simply selecting the perfect match. Instead, clinical and organizational factors combine to create physician preferences and organizational possibilities.

Factors that weigh in the decision to bring a patient into the ICU include not only patient characteristics, but also expectations of the quality of care here and elsewhere in the hospital, as well as the opportunity costs of committing the resources. The decision to discharge a patient from the ICU may include consideration of the expected future course of the patient, the costs of care, the consequences of inaccuracy in assessment, and the availability of alternative resources such as stepdown units, as well as the opportunity cost of keeping the patient in the ICU.

Admitting a patient to the ICU is a complicated business. Consider the effect of a *bounceback*. A *bounceback* is a patient who is discharged from the ICU, but because the patient's condition is not sufficiently stable on the ward, the patient returns to the ICU within hours or days. If, for example, the demand for ICU resources is acute and a patient is discharged from the ICU in order to use those ICU resources for another patient, a *bounceback* can create havoc. *Bounceback* also occurs at the floor level; an early discharge to home may result in a *bounceback* to the hospital. These events have specific costs and reimbursement consequences; a *bounceback* after discharge to home may not be reimbursed as a new admission but rather as a continuation of the prior admission.

In the hospital where the episode took place there were eight ICUs. Five of these were dedicated to the care of surgical patients. Of these five surgical ICUs, four were specialized (cardiac, burn, cancer, and neurosurgical services) and one was a general surgical unit. Although physically separate, for medical purposes, four of these ICUs (burn, cancer, neurosurgical, and general) were grouped together to act as a common pool of resources. [The cardiac surgical ICU was administratively separate and medically staffed by the cardiac surgeons and their residents who had no interaction with the other ICUs.]

More broadly, the appropriate use of ICUs is a cause celebre, and there are continual efforts to develop "hard" rules for deciding when a patient should be cared for in ICU. Despite these efforts, criteria for entry into and exit from ICUs remain difficult to define, vary between and even within institutions, and are rarely crisp.

At least some part of the daily need for surgical ICU beds can be anticipated. Certain surgical procedures, including many thoracic, cardiac, and neurosurgical procedures, routinely require post-operative care in the intensive care unit. These cases are usually scheduled well in advance of the day of surgery and the need for ICU beds is predictable.

The availability of beds in the ICUs is limited less by the physical number of beds than by the vicissitudes of nurse staffing; in this hospital there are more physical beds available than there are nurses to staff them. This was the case on the night of the episode and on most nights.

Most hospital patients reside not in ICUs but on floors that are each nominally associated with a particular surgical or medical service. There are neurosurgical floors, gynecological floors, etc. Patients coming out of a surgical ICU from a particular service are usually transferred to a corresponding floor. This is because the disease processes, patient characteristics, and postoperative problems that are most likely to be encountered are handled routinely on those floors. This is largely a matter of nursing expertise, which is refined over time to reflect particular kinds of patients. For example, patients who come from the neurosurgical unit require regular neurological checks to detect any change in neurological condition. Nurses on the neurosurgical floor perform many such neuro checks. Though routine on this floor, they are exceptional on other floors. Thus the strong preferences of surgeons for transferring their patients out of intensive care to "our floor" are quite rational. Also co-locating patients associated with a particular service aids the process of physician "rounds".

COPING WITH COMPLEXITY:
PARCELING OUT BEDS BY THE *BEDMEISTER*

In this hospital, the bed assignments for four units (burn, cancer, neurosurgical, and the general unit) are controlled by the *bedmeister*.[4] The person playing this role is responsible for managing the ICU resources, matching the anticipated load of patients and planning for daily admissions and discharges. Such a role is essential to any functioning ICU but it is critical in the setting of multiple ICUs with overlapping capacities, where the individual ICU resource pattern is only one part

---

[4] Bedmeister is a corruption of the German *bett* (bed) plus *meister* (master).

of the overall hospital ICU resource situation. Because demand for ICU "beds"[5] is only partly predictable, it is necessary to have a designated individual who is continuously aware of the resource situation and able to respond to new demands. The *bedmeister* during the day is the intensive care unit fellow.[6]

At night, two resident physicians are "on-call" to cover the four ICUs for general surgical patients. Because of the physical distribution of their respective ICUs, one is described as the "downstairs resident" (caring for patients in the general and cancer units) and one is the "upstairs resident" (caring for the burn unit and the general surgical portion of the neurosurgical unit). By convention, the downstairs resident is the nighttime *bedmeister*. Each of these residents is the primary physician responsible for the care of all patients in his or her units during the call period (typically from about 20:00:00 to 06:00:00). These residents are physically alone; specific issues may be referred to either the intensive care unit fellow or to the surgical service that is ultimately responsible for the care of the patient.

It is worth noting that the *bedmeister* role is a secondary one for the nighttime resident. The primary role of the nighttime resident is to care for the ten to thirty patients in the units, in itself quite often a demanding activity. Typically, the on-call resident works through the day and then throughout the call night without any sleep, but is the first to go home the following day—usually by 13:00:00 but occasionally as late as 18:00:00.

Artifacts as Tools: The Bed Book

Elective surgical procedures that will require a post-operative ICU "bed" should not be started unless intensive care unit resources are assured. To allow for the orderly accommodation of these needs, a bound notebook (known as the "Bed Book" or just *The Book*) is kept in the general intensive care unit and used for "bed" reservations for these patients. A reservation can be made at any time but, because of the way elective cases are scheduled, most reservations are made 24 to 72 hours in advance. *The Book* is public, and any surgical service can make a request at any time. The understood rule regarding "bed" requests is that they are honored on a first-come, first-served basis.

During the day, the ICU fellow and the night time *bedmeister* use *The Book* to anticipate the potential for future mismatch between the supply and demand for beds. For example, observing that tomorrow's surgical schedule will overrun the available resources (given the current occupancy and ICU admissions expected today) will lead to a making the discharge of a specific number of patients today a high priority.

---

[5] In this context, a "bed" does not mean the physical bed but rather the set of resources that are required for a patient; a "bed" is the smallest meaningful unit of ICU or ward resource. The term encompasses a collection of resources rather than a physical bed or patient cubicle.

[6] A "fellow" is usually a physician who has finished residency training and is doing an additional year of specialty training, in this case in surgical intensive care.

The incentives to discharge patients from the ICU in order to make room for new patients are not entirely aligned with the interests of the individual surgical service. Officially, any bed that is emptied by discharging the current occupant from the ICU is returned to the bed pool. This means that the bed is available to the *bedmeister* for assignment to any surgical patient, not only to patients from the service that made it available. This reduces the incentive to a surgical service to discharge their patients from the ICU; the resources may simply be consumed by some other surgical service. When a *bedcrunch* looms, however, the *bedmeister* can reduce the disincentive by agreeing to give the bed back to the discharging service for one of their new patients who needs an ICU bed. For example, if the day begins with full ICUs and no discharges to the floors are planned, the ICU fellow may refuse to allow elective surgical cases for a given service to proceed to surgery until that service has designated a bumpable patient from their service. Such arrangements tend to be ad hoc and politically sensitive for a variety of reasons.

The preceding has implied that ICU resources are homogeneous, i.e., that "a bed is a bed is a bed." This is not completely true: ICU resources are not entirely interchangeable. The assignment of patients to ICU beds takes place under an evolving set of restrictions. Some of these are strongly medically based. For example, a septic (infected) patient should not be placed in the burn unit because of the increased susceptibility of the other patients in that unit to infection. Some other restrictions are more a matter of surgeon preference. Certain surgeons, for example, will not allow their patients to be sent to certain units because they believe the care in those units is not comparable to those in their favorite unit. Even if this opinion cannot be objectively supported, the realities of political power make such restrictions de facto rules of placement.

In most cases, the decision to transfer a patient to the floor is made by the primary surgical service during the morning rounds. A resident on this service writes the transfer orders and the patient is transferred to the appropriate floor as soon as there is a bed available on that floor. The ICU cubicle is cleaned and made ready for the next patient, a process that may take as much as an hour. This fits well with the normal daily activity: patients are discharged from the ICU to the floor in the morning, the patient is moved and the ICU bed area cleaned and prepared for a new admission while the patient who will occupy that bed is undergoing surgery. Later in the day, the new patient comes to the ICU cubicle and the process continues.

Preparing for Demand > Supply Situations: Identifying *bumpable* patients

Because the hospital is a trauma center, and because patients sometimes become critically ill without warning, not all intensive care unit admissions can be planned. It sometimes happens that there is a demand for an ICU "bed" that cannot be immediately accommodated. In such situations it falls to the *bedmeister* to assess the patients in the ICUs and to select the least sick patient for transfer to a stepdown unit or the floor.

That this is possible at all is a function of the heterogeneous nature of ICU care and illness. Patients in the unit are in many different stages of illness and there are usually some who are nearly ready to be discharged from the ICU to the floor. These are candidates for earlier discharge in order to meet the immediate demands for ICU care of some other individual. In any event, discharge from an open ICU and transfer to the floor always requires the consent and active participation of the primary surgical service responsible for the patient.

When the ICUs are near capacity, the end of the day routine includes an informal conference between the intensive care unit fellow and the night's *bedmeister* to identify "*bumpable*" patients. A *bumpable* patient is one who is a candidate for transfer if an additional bed is required for a new patient. The advantage of identifying *bumpable* patients is that resolving new demands for ICU care is less likely to cause a major disruption and may be handled as routine. The notion of *bumpable* implicitly involves degraded levels of operations. The process of identifying *bumpable* patients is one means of planning for coping that characterizes many real world environments.

Son of Coping: Building an ICU from scratch when no ICU bed is available

If the units are full and ICU care for a new patient coming from the operating room is required, but there are no *bumpable* patients, a potential crisis exists that may require relatively extreme measures. Among these is the use of the post anesthesia recovery room (PACU) or the operating room (OR) itself as an intensive care unit.

These approaches are expensive, disruptive, and severely limited. Using the PACU or *operating* room as an ICU utilizes nurses who are not generally involved in ICU care to provide ICU-like resources. The normal ICU activities cannot encompass these settings. The per hour cost of PACU and OR are on the order of 4 to 10 times as much as that of the normal ICU.

Because these areas are physically isolated from the ICU, policy requires that a *surgical* resident stay with such a patient continuously until the patient can be transferred to a genuine ICU. This means that the surgical resident is unavailable for other work, say, in the emergency room or on the floor.

In addition, consuming the operating room or PACU for ICU-like use makes them unavailable for other work. At night, when there is a limited number of OR teams available to do cases, tying those teams up as ICU personnel means that the ability to handle new OR cases is diminished.

Piling Pelion on Ossa: Escalating demands

At night, the *bedmeister* function is only one part of the "downstairs" resident's job. The need for the bedmeister function increases as the other work of the resident increases. When the ICUs are not full and there is a new demand for a "bed", the *bedmeister's* job is easy: it is only a matter of deciding to which of four units the new patient should be sent. But in this situation, the *bedmeister* in role of caregiver is relatively untaxed; because the census is low and "beds" are available, the resident confronts comparatively low workload. In contrast, if the ICUs are full, the *bedmeister* function becomes more difficult and time consuming; at the same time, the number of patients requiring care is maximal. Thus the effort demanded to resolve requests for ICU resources is highly correlated with the other demands for resident time and attention.

Given these insights, read the simple story again:

A patient undergoing surgery needed to go to an intensive care unit when the surgical procedure was complete. The surgical procedure was finished before the intensive care unit bed was available, so the patient was taken first to the recovery room and transferred to the intensive care unit after a few hours.

Now look at how it actually unfolded:

Table 1. Critical Incident Sequence

Background: The episode occurred in a month when the ICUs were continuously working at capacity; the demands for "beds" were intense. For several days including the night of the episode, the units were full. Beds were available for new patients only as old patients were discharged.

Sequence of Events:
1. Surgical service "A" fails to reserve a bed in the open intensive care unit for a surgical case scheduled for the late afternoon and the case is placed on "hold" because no beds are available.
2. The ICU fellow OK's the beginning of the surgical case, knowing that there are one or two bumpable patients in the unit. The fellow instructs the *bedmeister* to identify and bump a patient.
3. The *bedmeister* identifies the least sick patient in the unit. This patient is a trauma victim who had been admitted 24 hours earlier and has had a relatively stable course. Because the ICU is "open" the *bedmeister* calls the surgical service caring for the trauma patient and requests permission to transfer that patient to the regular floor. The *bedmeister* also asks the ICU clerk to call the admitting office (which clears requests for beds on the floors) and obtain a floor bed assignment.
4. The surgery service declines to transfer the trauma patient but instead recommends another post-surgical patient for discharge from the ICU and transfer to the floor. This patient has had relatively minor surgery but is undergoing a ROMI or Rule Out Myocardial Infarction (heart attack) procedure and so must be transferred out not to a regular floor but to a telemetry bed for continuous ECG monitoring. The *bedmeister* notifies the ICU nurses and ICU clerk of this change in plan.
5. The patient selected for transfer has a pulmonary artery catheter. These catheters are permitted only in intensive care units and the catheter must be removed and replaced with an intravenous line. The *bedmeister* changes the pulmonary artery catheter to a triple lumen catheter in a sterile procedure.
6. Two hours pass. The *bedmeister* attends to other duties.
7. The surgical procedure is proceeding smoothly and the circulating nurse in the operating room calls to see if the ICU bed is ready. The *bedmeister* notes a new laboratory value showing that the patient to be bumped now has a blood sugar of 450 mg/dl (high) and blood and acetone in the urine.

The *bedmeister* concludes that the patient is in diabetic ketoacidosis (DKA) and cannot be transferred to the floor. The *bedmeister* cancels the transfer and begins treatment of the DKA.

8. The *bedmeister* identifies another *bumpable* patient for transfer. This patient had a severe nosebleed the night before, and required multiple blood transfusions. The patient has not had any bleeding for 18 hours. The *bedmeister* contacts the surgical service caring for this patient and obtains permission to transfer this patient to the floor. The patient does not require continuous ECG monitoring and so cannot be transferred to the telemetry stepdown unit but must be transferred to a regular floor. The *bedmeister* asks the clerk to contact the admitting office to clear a floor bed for this patient. The *bedmeister* discusses the situation with the ICU nurse in charge. They elect to "hold on" to the telemetry bed (and not to notify the admitting office that the telemetry bed is now not required) for possible use in the event of another request for an ICU bed, assuming that the first patient will again be *bumpable* at that time.

9. The ICU nurses notify the operating room nurse that the ICU bed will not be available for at least 1 to 2 hours. The transfer to the floor will delay the bed availability.

10. A surgical resident is asked to come to the ICU to write transfer orders for the patient being bumped. The resident comes to the ICU and writes the transfer orders.

11. The OR nurse foresees that the surgical procedure will be finished well before the ICU bed is available and notifies the operating room desk nurse who is instructed to alert the post anesthesia care unit (PACU) nurses. The PACU nurses normally leave the hospital after the last ordinary (non-ICU bound) operating room case is completed; patients going to the ICU are transferred there directly without first going to the PACU. In this case, the PACU will be used as a temporary intensive care unit and the PACU nurses will need to stay in order to fill the gap between the end of the surgical procedure and the ICU bed becoming available.

12. Because the patient will require a mechanical ventilator, something not normally used in the PACU, the PACU nurses page the respiratory therapist for a ventilator for the patient.

13. Elsewhere in the hospital, a bed is assigned to the bumped patient from the ICU and nurses prepare to receive that patient. ICU nurses prepare to transfer that patient. Cleaning personnel prepare to clean the ICU location as soon as it is vacated.

14. In another operating room, an unrelated *cardiac* surgical procedure is nearly finished. The circulating nurse in *that* room calls the cardiac surgical ICU and requests delivery of an ICU bed to the room. The cardiac surgical ICU nurses call for a "patient care technician" (a low skill hospital employee whose job includes transporting equipment) to move the bed from the cardiac surgical ICU to the operating room corridor.

15. A patient care technician picks up and delivers an ICU bed not to the area outside the cardiac surgical operating room but rather around the corner outside the operating room in which the episode case is proceeding. The technician is a relatively new employee and fails to recognize that the bed is placed in a non-standard position, closer to the room in which the episode case is going on than to the adjacent cardiac surgical suite where the patient for whom the bed is intended is undergoing heart surgery.

16. A nurse in the episode case operating room sees the ICU bed in the hallway outside the operating room. Because ICU beds are brought directly from the ICU cubicle to which they will return, this is interpreted by the nurse as a signal that the ICU is now prepared to receive the episode patient. The nurse concludes that the ICU bed is available and notifies the PACU nurses that they do not need to remain because the patient from the episode case can be taken directly to the ICU.
17. The PACU nurses notify the respiratory therapist on-call that the ventilator is no longer required.
18. The ventilator is removed from the recovery room. The PACU nurses prepare to leave the hospital.
19. The episode case is completed and the preparations are made to move the patient onto the ICU bed.
20. The *bedmeister* stops by the operating room to see how the case is going. The circulating nurse for the operating room comments that the patient is ready for transport to the ICU. The *bedmeister* tells the circulating nurse that the ICU bed is not yet available and that the plan to transport the patient to the recovery room has not been changed. The circulating nurse and the *bedmeister* exchange harsh words.
21. The *bedmeister* goes to the PACU and intercepts the PACU nurses as they are preparing to leave. They are informed of the fact that the original plan remains in force and ordered begin to prepare for the transport of the patient to them. More harsh words are exchanged.
22. The episode case patient's transfer is put on hold in the operating room pending the delivery of a ventilator to the recovery room, thus turning the operating room into a temporary ICU.
23. A ventilator is delivered to the PACU.
24. The episode case patient is transported to the PACU. Report of the case is given to the PACU nurses.
25. The floor nurse notifies the ICU nurses that a bed is now available for the *bumpable* patient. This patient is transferred to the floor.
26. The ICU cubicle is cleaned and prepared for the episode case patient.
27. The episode case patient is transported from the PACU to the ICU. Report of the case is given by the PACU nurses to the ICU nurses.
28. The next morning the *bedmeister* describes the sequence of events to the administrative head nurse of the ICUs. The nurse identifies the root cause of the difficulty as one of "human error" on the part of the patient care technician who delivered the ICU bed to the wrong hallway location.

Aftermath:
The patient in this episode suffered no direct harm. A partial list of the consequences of this failure though includes the following:

• Extra operating room time and the time of two nurses and three physicians (surgical resident, anesthesiology resident, anesthesiology attending staff).
• Extra recovery room time including the time of nurses and respiratory therapists; double setup of a ventilator.
• Multiple exchanges of responsibility for the patient's care with resulting information loss about the conduct of the case and the patient's condition.
• Time and effort to schedule and reschedule beds. The loss of telemetry resources from the larger pool for one evening.
• Extra time spent by the *bedmeister* away from the patients in the ICU.

OBSERVATIONS ON THE INCIDENT

The administrative nurse's attribution of the failure to "human error" on the part of the patient care technician is remarkable. The focus on the most proximate flaw in a long series of flaws and faults as the "cause" is the usual circumstance in complex system failures (Hollnagel, 1993; Woods, et al., 1994).

Interestingly, the event occasioned no deeper reflection on the part of the administrator. No thought was given to the nature of ICU resource use or scheduling, the roles of various parties in resolving resource conflicts, the impact of operating a hospital at or near its theoretical capacity, the potential consequences of the extra activities required of the many people in the ICU and on the floor to cope with these circumstances, or the implications of the "hiding" of the (unused) telemetry bed. What is remarkable about this episode is that the events were normal and even mundane.

The sequence appears almost Byzantine in complexity. Yet even this telling of the story is a pared down account derived from pages of notes. In fact, I argue that all accidents in medicine have similarly complicated stories that might be told but generally are not (e.g., Cook et al., 1998). The decision to tell a longer or shorter version is arbitrary. Reading and understanding this episode, even simplified as it is, is a demanding task. More significantly, the detailed recounting of episodes requires both technical knowledge (e.g., what does diabetic ketoacidosis mean? what is a pulmonary artery catheter?) and organizational knowledge (e.g., how are ICU assignments made?) for comprehension.

Comprehension also requires understanding of the local factors and roles (e.g. the *bedmeister*) that may differ substantially between hospitals and in a given hospital over time. As the story gets longer, the background knowledge required becomes ever more complicated. The present episode demonstrates the great difficulty in communicating succinctly the details of the sorts of events that permeate everyday practice. It also demonstrates the complex interconnectedness between domain knowledge and practice that is often missed or poorly examined during investigations of accidents.

The episode also calls into question commonplace views of medical knowledge as the ground substance of clinical decision making. There is little in the description about the underpinnings of medicine and nearly nothing about classical "medical decision making." While all of these components are present in the work and the participants could describe them if questioned, they remain submerged within the activity of the parties involved. This is in keeping with the inner stories of many accidents where the participants' activities and expressions focus on a broad set of pragmatic issues regarding the system in operation. The crucial domain specific knowledge (e.g., what constitutes an unstable patient, what medical details are relevant in determining whether a patient is *bumpable*) are embedded in the workaday world knowledge of the participants.

Practitioner argot is often regarded as simple shorthand for specific technical terms in the language of the domain. But the suggestion here is that argot performs a more complex function than simply standing in, one for one, for technical terms. The practitioner language provides useful aggregates of domain knowledge. These aggregates refer to factors within and across the levels of a goal-means abstraction hierarchy (Rasmussen, 1986). Being *bumpable* means having temporarily taken on a certain characteristic that has individual and larger group meanings. Medical stability, the threat of future clinical events, the ease with which these may be detected, the consequences of failing to detect them, all play into the assignment of this state. But being *bumpable* is more; it depends not only on inherent characteristics of the patient but on the states of other patients, the needs of other units, the other available resources. Being *bumpable* involves comparisons, contrasts, and expectations. It is not a static circumscribed technical term, like "coronary artery disease" or "diabetic ketoacidosis," but a fluid, dynamic, complex, time limited, assessment—characteristics that it shares with the domain. It subsumes a variety of local factors, local issues, local limits and local capabilities.

The use of the term "*bumpable*" is not ephemeral or insubstantial; it has real meaning that is subject to negotiation between practitioners and throughout the organization. The argot is a powerful tool used to express the relationships between higher and lower abstraction levels, neither just technical nor just organizational but rather both. That practitioners develop and employ such argot indicates that their work is not divided between technical and organizational elements but is instead about the interplay between these.

Bumpable has at least one other important quality in this context, namely its forward projection. *Bumpable* refers to a present state with future consequences. It is inherently about planning for a potential future, of anticipating and preparing to cope with unknown and unknowable future events. It references a hypothetical future time in which the practitioners will have to deal with a degraded system. It shows one means by which practitioners reach forward into a hypothetical, conflicted, demanding future. By declaring patients *bumpable*, practitioners envision a future world which will demand that they act in ways that push the system towards saturation and, potentially, towards overt failure.

Galison (1997) has pointed out the presence of specialized "trading languages" in the high-energy physics community. These allow individuals from different disciplines (e.g., refrigerating engineers, experimental physicists, theorists) to exchange information about the interacting characteristics of their work. The argot of intensive care practitioners is an instance of this. *Bumpable* allows nurses, surgeons, administrators, and cleaning people to exchange information about a particular patient, to declare intentions, to argue about the directed meanings of technical and organizational factors, to consider the future pace and character of work, and to foresee hazards. *Noto bene* that argot terms do not and cannot have a universal meaning for everyone. For a nurse, *bumpable* implies, perhaps, preparation of papers and communications for transfer. For the nurses' aide it implies a future empty bed that will need to be made ready for a patient. For the resident it implies a resource to be consumed at a future time. The term provides both discipline specific meanings and simultaneously a shared representation that allows work to be organized and communication to proceed.

The very presence of these terms is a clue to the difficult, demanding, and hazardous locales within the domain of practice. The notion of *bumpable* is directed at the **margin of practice** rather than at the median of the distribution. Because this system typically runs at or near capacity, it is the marginal events and circumstances that are the focus of attention, that threaten to saturate resources, that generate disputes, rancor and confusion, and that elicit expertise. There are other terms (e.g., *bounceback*, *bedcrunch*) with similar characteristics. Argot thus plays a central role in the planning to react and the reacting by replanning that characterizes this and other domains, and this plan/reaction is directed not at the system's center of mass but at the edges where failure and success hang in the balance.

The suggestion here is that practitioner argot points directly to the critical concerns of NDM researchers. Because they span the distance from technical to organizational, these terms capture the essence of what is going on and therefore are at the locus of the decisions/actions of practitioners. To explore what it means to be *bumpable*, what a *bedmeister* actually does, what a *bedcrunch* entails, what *The Book* actually does, is to explore the core of success, failure, coordination and expertise.

Chapter 4

Discovery as Functional Synthesis

Doing a cognitive task analysis is like trying to learn the secret of a magic trick: once you know the trick and know what to look for, you see things that you did not notice when you did not know exactly what to look for.

Paraphrase of Orville Wright's comment about discovering the secret of flight (Woods, 2003)

"BEING *BUMPABLE*" AS AN EXAMPLE OF STUDYING A JCS AT WORK

The synthesis of cognitive system functions in the ICU was built around the practitioner term *bumpable*, and it reveals much about discovering patterns in JCSs at work. To tell the story of the critical incident and put it in context, the account of how ICUs function is full of details, described in domain terms. But the selection of what details to notice and how to group them into a coherent description presupposes a sensitivity to model the ICU as a Joint Cognitive System (JCS) and to recognize what ICUs adapt around, e.g., *bounceback, bedcrunches*. The incident and the larger account of ICUs illustrates Ashby's Law of Requisite Variety in action (*JCS-Foundations*, p. 40).

Note that adaptations often become routinized as a standard part of a task or a role, so that, on the surface, it is difficult to see how these routines are adaptive and to what they have adapted (cf., the Law of Fluency, p. 20). Similarly, the adaptations as exercised in everyday practice are not necessarily noteworthy to the practitioners themselves. Their ability to tell people directly about these processes is limited, since there is usually a significant gap between what people say they do and what they are observed to do unless special procedures are used. The synthesis of cognitive system functions identifies what behavior is adapted to; yet some of these factors dominate the account of how the JCS works well out of proportion to their frequency of occurrence. This means that discovery is aided by looking at situations that are near the margins of practice and when resource saturation is threatened (attention, workload, etc.). These are the circumstances when one can see how the system stretches to accommodate new demands, and the sources of resilience that usually bridge gaps.

Notice how practitioner argot or terms of practice provide clues about how to meet the challenge of the Law of Fluency in order to discover what the system is adapted to, and how. These terms often point to informative relationships, integrative judgments, coordinative links, and the responsibilities attached to roles that reveal the nature of practice and that are missing from or are under-emphasized in formal accounts of operations.

Critical incidents recognized and analyzed as breakdowns or near-breakdowns in adaptation help in the discovery process, and are useful heuristics for organizing and communicating the results across series of investigations that look into how the JCS functions. Recognizing an episode as a story of adaptation, resilience, and brittleness, when it is otherwise not noteworthy or reflected on by participants, is a key skill to develop as an observer of JCS at work.

General patterns in JCSs at work also run throughout Cook's account of the ICU. One begins to see the ICU as part of a larger dynamic process where patients enter and leave as their physiological status degrades or improves. The ICU is more than simply a place for care. Issues of responsibility and authority accrue around the roles of different participants in the care process (e.g., getting a patient to the appropriate unit of care). The processes of workload management are central to the description as we hear of adaptations intended to cope with or avoid workload bottlenecks (recruiting resources to handle workload surges, moving or chunking tasks in time to reduce workload bottlenecks or competition). Nevertheless, occasionally, as in the incident Cook captured, crunches occur.

Anticipation runs through the discussion of the *bedmeister* role (Klein et al., 1995). Cook's synthesis shows how a resident, when effective in the *bedmeister* role, develops a readiness to respond (Hollnagel & Woods, 2006)—when will events combine to challenge ICU capacity? Are current ICU patients improving, and at what point might they be moved or discharged? The functional analysis also captures how events cascade and demands escalate, putting a squeeze on the person in the *bedmeister* role given their other roles in patient care. These cascades can produce multiple interacting demands, illustrating a general pattern in JCSs. Good syntheses of how a JCS functions find these cascades and squeezes, and how practitioners cope in the particular setting of interest (Woods & Patterson, 2000).

To recognize that there are states like *bumpable*, roles like *bedmeister*, potential bottlenecks such as a *bedcrunch*, and practitioner created tools like *The Book,* is to engage in the process of **discovery** central to the study of JCSs at work. By taking up this challenge, one can propose a functional account of the setting as a JCS, the way this system adapts to achieve success, its vulnerabilities to breakdown, the constraints that drive adaptations, and the meaning of what it is to practice.

INSIGHT AND FUNCTIONAL SYNTHESIS

If the intensive care unit case of synthesizing functions in a JCS ("Being *Bumpable*") is a model for research and design, then apparently we are all left in a quandary. Where are the methods? Look at two more classic cases that synthesize an account of a JCS and how it works: Ed Hutchins' study (1995b) of a commercial transport cockpit as a JCS—"How a cockpit remembers its speeds"—and the account of how voice loops serve as the key coordination tool in space shuttle mission control, by Patterson, Watts-Perotti & Woods (1999).

The authors in each case present insightful syntheses of how these control centers are "well"-adapted (the relationships of the artifacts, the demands, the coordination strategies) based on extensive access to these fields of practice, and detailed *observation* of practitioner activities in context. But again, the astute reader concerned with how to do such studies will be left adrift: what is striking in each paper is the virtual absence of any discussion of methods. Yet this trio of cognitive task syntheses are wonderful examples of functional modeling, and researchers concerned with cognitive systems and work should aspire to reach, at least, the standard set by "Being *Bumpable*" and these two other studies.

Three points are striking about this set of studies: (1) the power of observation, (2) general patterns expressed in particular situations, and (3) a proposed functional account. First, the trio of studies captures the power of observation. While the studies build on intimate connection to the field of practice (authenticity), i.e., to be able to observe from inside, the power of observation lies in how the investigators see beyond the surface and provide a fresh, surprising account. How does one learn to see anew like this? There are no routines that automatically produce such skill, nor are there procedures that guarantee the desired end result. As is true for all skills, skill at observation must be practiced and supported in order to grow.

Let us provide a few observations about "observation" as illustrated in "Being *Bumpable*" and the other two studies. One maxim for skill at observation is: *appreciate the significance of small details*; though the catch is: *not all small details are significant*. Each study began with noticing the significance of what might be seen as just another detail: the practitioner term "bumpable" used in ICUs, the demand on the pilots to remember different speeds during a descent into an airport, how space mission controllers listen in on voice communication channels. Notice how most inspections of these work settings would regard these as of little value: a piece of jargon that blocks communication, a detail to be remembered which should be automated away in more advanced technology, a noisy antiquated communication mechanism that should be eliminated. Yet the investigators used each starting point as a wedge to open up the adaptive process, reveal the nature of expertise, and identify basic demands/constraints that must be supported.

Each study also began with observers who were *prepared to be surprised*. This succinct phrase captures so much about the heart of science as well as about the process of observation. What forms of preparation are needed so that one can notice what they never expected to see? When observation began in each setting (ICU, cockpit, mission control center), none of the observers could articulate a detailed hypothesis to be disconfirmed, yet each recognized what was not obvious and pursued that lead. In part, the preparation comes from using a pattern base built up through previous observation and abstraction across apparently diverse settings. This preparation allows the observer to constantly ask: What is this activity an example of? Have we seen this before? How does this both fit and differ from patterns in the research base?

The role of preparation for surprise in the process observation is another case of Neisser's (1976) perceptual cycle where there is an interplay between calling to mind general knowledge about JCSs at work and observing specific JCSs working. Hogarth (1986, p. 445) used the following quip to capture the point: "One does not send novices to the North Pole or to the moon since they would not know what to look for." The skill is a general one not only in studying JCSs, but in science and in diagnosis—using concepts to see beyond the surface variability without being trapped into seeing only those concepts.

Second, each of the studies derives patterns to be accounted for. They note deeper aspects of the field of practice, which are in need of explanation in a functional account. "Being *Bumpable*" identifies patterns that are consistent with concepts for workload management, control of attention, and others. These patterns guide further observation by specifying how to shape the conditions in order to better see more deeply into the processes or dynamics in question. The warrant for these patterns comes from several sources: (a) claims to authenticity in the set of observations and (b) matches (or mismatches) to the patterns noted in previous studies of this setting and to the joint cognitive system themes invoked. Is it confirming, expanding, modifying or overturning these generic patterns?

Third, each study provides a coherent account of the field of practice that goes behind the veil of surface characteristics to propose deeper dynamics. Providing an account or candidate functional explanation about how behavior is adapted to constraints and goals makes a sufficiency claim relative to the sets of observations (does the account cover the observations?) and relative to other functional accounts previously offered (does it cover the observations better, more simply, more widely?).

These three characteristics of syntheses of how joint cognitive systems function presuppose a shared cross-investigator pattern base. CSE requires a means to share observations and patterns from various field studies among a community of researchers interested in those cognitive system themes across diverse settings. It also requires a means to share functional models among the communities of researchers and designers of JCSs in general and in particular settings. Sharing observations, patterns, and functional accounts does not imply agreement or simple accumulation, but rather, sharing refers only to making the accounts and their basis in observations explicit and accessible so that we can contrast patterns to develop new accounts and challenge explanations through new observations.

Ultimately, what is striking in the trio of studies as functional accounts, is insight. Insight generates ideas which stand on their own and which are open to argument and revision through new observations and alternative covering accounts. Insight as a discovery process is measured by models of argument (such as Toulmin's (1984) structure of argument (Toulmin et al., 1984) or accounts of abductive explanation as a model of justification in argument), not by the syntax of inductive or deductive method. In other words, insight provides the linkage between field research techniques as discovery processes and laboratory research techniques as verification processes—hence, Bartlett's classic dictum in psychology (1932) to begin experimental studies with observations of ongoing activities *in situ*.

Insight is valuable; indeed, it is primary, in itself, for science. From a discovery point of view, insight has no need of justification by pure method (Woods, 2003). It does not demand repeatability or indifference to the investigator or investigative team (what some criticize erroneously as "subjectivity" in field research). Ironically, almost universally, discovery continues to be treated as ineffable in accounts of science and in teaching of research methods (attempts have begun in cognitive science to study and model discovery processes, but even these limited advances have made little impact on science training and education). Yet all programs of research in science embed processes of insight and methods to facilitate such insights; the descriptions of research process that are socially approved in communities of researchers generally ignore processes related to insight. The absence of, or even taboos on, discussion of insight and discovery do not facilitate these essential processes. As goes for any and all aspects of work, while there are individual differences in acquiring and demonstrating skill, processes for discovery are train-able and aid-able. One can view the techniques for functional synthesis and field research discussed here and elsewhere as a powerful and pragmatic program for facilitating insight (as noted by Neisser in the epigraph to Chapter 2).

How does one take up functional synthesis as a research program? The starting point is to orchestrate a variety of ways for shaping how one observes JCSs at work.

Chapter 5

Shaping the Conditions of Observation

The laws of demands and fluency capture the significant difficulties that one faces in studying JCSs at work. How does one break through the current web of adaptations to understand the underlying demands and to anticipate/guide future processes of change and adaptation? Without this understanding processes of change are exercises in slow trial and error despite the extravagant claims of technology advocates (Woods & Christoffersen, 2002).

In general, effective methods for studying JCSs at work disrupt and disturb the current system to observe how that system adapts and settles back into equilibrium. This basic technique for functional analysis can be carried out in multiple ways:

(a) Observing at naturally occurring points of change; for example, see the series of studies of the side effects of introducing new information technology into high performance health care settings such as Cook & Woods (1996), Patterson et al. (2002), Ash et al. (2004), Koppel et al. (2005); Wears & Berg (2005).

(b) Reconstructing the run up to accidents and incidents since these breakdowns reveal the adaptive limits of the JCS (e.g., critical incident methods such as Klein's techniques for eliciting stories of challenge and response from practitioners; see Klein et al., 1989; Hoffman et al., 1998; Klein, 1999; or process tracing in accident investigation; see Dekker, 2002).

(c) Introducing probe events at the suspected boundaries of the current system to force adaptive behavior into the open; this is well-illustrated by the scenarios designed for the full scope simulation studies of pilot interaction with cockpit automation (see Sarter & Woods, 2000).

(d) Changing the artifacts available for use in handling situations; the contrast in process and performance as artifacts are changed is nicely done in Sarter's studies of tactile indicators of indirect mode changes in pilot interaction with cockpit automation (Sklar & Sarter, 1999), and in a study of how experienced users avoid getting lost in complex human-computer interfaces (Watts-Perotti & Woods, 1999).

The first two classes of methods for studying JCSs at work are natural history techniques—a diverse collection of observations *in situ*. The second two classes of methods are staged or scaled world techniques, and critically depend on having problems designed to present hypothesized demands. Note that natural history techniques help design problems by identifying situations that present demands and dilemmas. The warrant for results from staged world studies is increased by the ability to link the probes introduced through problem design back to actual incidents and events in the field of practice.

THREE FAMILIES OF METHODS THAT VARY IN SHAPING CONDITIONS OF OBSERVATION

When one sees research methods as variations in shaping the conditions of observation, three classes of research methods appear along this base dimension (Figure 5):

• Natural History methods (*in situ*)
• Experiments-in-the field: Staged or Scaled Worlds
• Spartan lab experiments: Experimenter-created artificial tasks.

These three classes are distinguishable by determining the "laboratory" used. Natural History techniques are based on a diverse collection of observations *in situ* (DeKeyser, 1992; Hutchins, 1995a; Fetterman, 1989). Experiments in the field begin with Scaled World simulations that capture or "stage" what is believed to be the critical, deeper aspects of the situations of interest from the field (e.g., De Keyser & Samercay, 1998). Spartan Lab techniques focus in on a few variables of interest and their interactions in experimenter created (and, therefore, artificial) situations.

Note the use of the words "weak" and "strong" to describe the vertical dimension in Figure 5. Weak shaping of the conditions of observation is both a strength and a weakness; strong shaping of the conditions of observation is both a weakness and a strength, though each in complementary aspects. Hence there is a need to orchestrate varieties of observation that are diverse in how they shape (and therefore distort) the processes we wish to observe and understand.

Natural History techniques are based on a diverse collection of particular observations, "which agree in the same nature, though in substances the most unlike" as Bacon put it at the dawn of science (1620). In effect, the natural history observer is listening to, eliciting, and collecting stories of practice. Building up a corpus of cases is fundamental (e.g., see Sarter & Woods, 1997, for a corpus of cases, how results from previous studies shaped the collection process, and how the corpus contributed to a subsequent staged world study in Sarter & Woods, 2000).

Natural History begins with the analysis of the structure or process of each case observed. The results are then compared and contrasted across other analyzed cases (requiring observations across various conditions). Observation *in situ* is jump started by reference to the pattern base as one asks: What patterns and themes in JCSs at work appear to play out in this setting? What kind of natural laboratory is this or could this be?

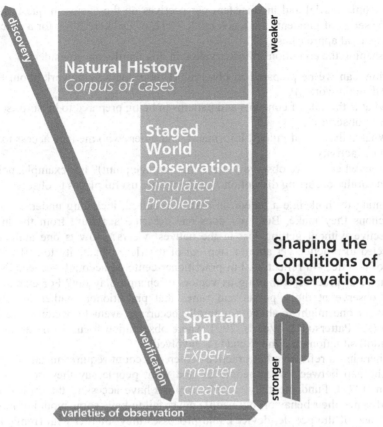

Figure 5. Three families of variants on shaping the conditions of observation.

In Natural History methods, a critical constraint is that observers have to wait for what happens; however, what happens is authentic by definition, though only a sample. By shaping the conditions of observation, we can seek to increase the chances or frequency of encounter of the themes of interest. For example, one might decide to observe a crossroads setting (e.g., the intensive care unity in medicine) where many diverse activities, tempos, and agents interact at different times, then use patterns noted to choose a tighter focus for future rounds of observation.

Very often, eliciting and finding stories of challenge/response in practice is a retrospective process in that the stories collected have already occurred. The limits of retrospective accounts are balanced when one builds a corpus of related cases and when one is able to collect multiple participants' accounts of the challenge/response process for that case. Anyone who has been part of an accident or critical incident investigation understands how different participants' accounts provide only partial and inconsistent perspectives on the events in question (for a specific series of incidents see Cook et al., 1992, or Dekker, 2002, for a description of the general approaches).

In shaping the conditions of observation in this family, one considers:

• How can we be prepared to observe? (Doing one's homework about how the domain works.)
• What is the role of concepts and patterns in being prepared to be surprised?
• Who observes?
• What is the role of guides, informants, and others who mediate access to the field of activity?
• From where do we observe (what is the vantage point)? For example, points of naturally occurring disruption and change are useful places to observe.

Coming to think like a native, in part, is required, including understanding the distinctions they make. But how does one describe a culture from the insider's perspective without getting lost in the natives' views? How is one authentically connected to practice yet able to step out of the flow of activity to reflect on the heart of practice—a reflective, but practitioner-centered account—especially when the nature of practice is moving as vectors of change play out? For example, one might observe at those points and times that practitioners gather in reflective forums, or one might re-shape a naturally occurring event to become a reflective forum (see Patterson & Woods, 2001 where observation focuses on various types of handoff situations and the exchanges involved).

Generating a reflective but practice-centered account requires means for coping with the gap between what people do and what people say they do (Nisbett & Wilson, 1977). Fundamentally, people may not have access to the critical factors that influence their behavior, but people never fail to have some model of how they work—and of the people, devices, and processes they interact with (reminding us that there is no such thing as a cognitive vacuum for people as meaning-seeking agents, see the law stated on p. 11). This also is related to the problem of *referential transparency*, as ethnographers call it (Hutchins, 1995a), which refers to the difficulty in seeing what one sees with.

Staged World observation falls in between Natural History techniques and Spartan Lab techniques (Woods, 1993). This family is often viewed as a weak or limited version of real experiments, degraded and limited by the context (and by implication, the skill of the investigator). Rather, this area on the dimension of shaping conditions of observation is a unique, substantive family of methods that is different from lab experiments and natural history, though it overlaps with each in several ways (see as examples, Roth et al., 1987; Layton et al., 1994; Nyssen & De Keyser, 1998, Johnson et al., 2001; Patterson et al., 2001; Dominguez et al., 2004; Norros, 2004; Clancey, 2006).

The key in this family of methods is that the investigators stage situations of interest through simulations of some type. The concrete situation staged (the test situation where observation occurs) is an instantiation of the generic situations of interest (the target class of situations to be understood). The mapping between the target situation to be understood and the test situation where observations are made must be explicit, since it represents a hypothesis about what is important to the phenomena of interest. Classically this is called the *problem of the effective stimulus* (out of all of the aspects present in the situation and all of the different ways one can model the situation present, which ones are people sensitive to and influence their behavior; see Duchon & Warren, 2002 for an excellent example). Thus, the fundamental attribute to Staged World studies is the investigators' ability to design the scenarios that the participants face. Problem sampling is the focus of this family of techniques.

Staging situations introduces a source of control relative to Natural History methods as it allows for repeated observations. Investigators can seize on this power to design scenarios and contrast performance across inter-related sets of scenarios—if something is known about what makes situations difficult and about the problem space of that domain. Though it must be remembered that while one may try to stage the situation the same way each time, each problem-solving episode is partially unique, since some variability in how each episode is handled always occurs.

Second, the power of staging situations is that investigators introduce disturbances or probes that instantiate general demands in domain-specific and scenario-specific forms. This allows for observation of how a JCS responds to the challenges embodied by the probe in context. The observer traces the process of response as the JCS moves toward a resolution. Probes are fashioned as complicating factors that move scenarios beyond "textbook" cases (Roth et al., 1987; Roth et al., 1992).

The capability critical for moving to Staged World studies is this ability to *design problems* and then express these problems in specific scenarios (by the way, this is another illustration of Neisser's perceptual cycle in action where the cycle now applies to a part of the research process). A central skill in CSE (arguably *the* central skill) is this ability to design problems and instantiate them in scenarios meaningful to actual practitioners. Scenarios, then, are not just interesting or difficult cases in themselves. Rather they have a *parallel* status as a tentative model of practice—what is the envelope of textbook situations, what complicating factors challenge this envelope, what demands and constraints drive practice. This means that the investigator works through the target test mapping explicitly and in detail. In designing the situation, probes, and context, the goal is to uncover or make tangible, and, therefore, observable, what has been encoded in practice (fluency law).

Detailed descriptions of how to design problems and their instantiations in specific studies are in Woods et al., (1990) & Schoenwald et al., (2005). A test for the reader would be to re-examine Cook's "Being *Bumpable*" functional synthesis of intensive care units, and begin to design problems expressed in terms of challenges such as *bedcrunches* which could be used in Staged World studies.

Once the problem structure is defined relative to the domain-specific characteristics, one can then trace the process of how the JCS responds to the challenges embedded through the probe events in context. For example, how does the presence of an initial disturbance that masks key cues to the main fault affect the diagnostic search process? This is why the analysis methods are referred to as "process tracing" methods (Woods, 1993).

Process tracing methods follow the steps specified by Hollnagel et al. (1981) as sketched in Figure 6 and as summarized in Table 2 (see Roth et al., 1987 or Nikolic & Sarter, in press, for a process tracing analysis in full). The starting point is analysis of the process for handling the situation in its specific context (the description of how the scenario was handled in terms of that world of work). Second, the general patterns or demands about JCSs that were instantiated in the concrete case are used to shift the analysis away from the language of the domain and specific scenario, and into an abstract story in terms of the concepts being studied.

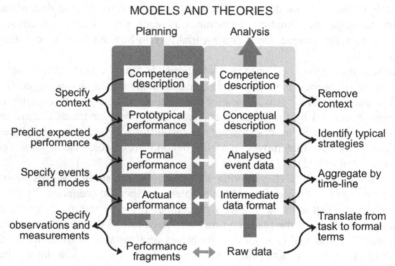

Figure 6. Levels of analysis in process tracing studies (Originally from Hollnagel et al., 1981).

The analysis then shifts to develop a description of how practitioners handled the demands of the problem designed into the situation staged (formal descriptions of performance). At this level of analysis, a study re-tells each episode in terms of general conceptual patterns—how over-simplifications led to erroneous actions or how the participants avoided this vulnerability (Feltovich et al., 1997), or, in another study, how detecting deception is difficult (Johnson et al., 2001), or how people cope with the brittleness of expert machine advisory systems when complicating factors arise (Roth et al., 1987; Guerlain et al., 1996). For example, if one's investigation concerned Feltovich's concepts about over-simplification tendencies (Table 4; Feltovich et al., 1997), then these concepts provide a set of story archetypes that are used to provide a more general description of the behavior of practitioners relative to the problem presented—e.g., observing how the situations created opportunities for people to be trapped in over-simplifying complex processes and understanding what helps or hinders people when they need to shift from simpler to more complete models of complex processes in order to handle anomalies (Feltovich et al., 2001).

Third, one contrasts different formal performance descriptions with different contrast cases—across different but related probes, different scenarios, domains, or artifacts. These various cross-protocol contrasts help the researcher abstract patterns and build possible functional accounts of what is competent performance in the face of demands and how performance is vulnerable to breakdown (e.g., the patterns abstracted across protocols and domains in Woods & Patterson, 2000).

Table 2. A Checklist for Designing Staged World Field Studies (Woods 2003)

STEPS	PROBLEMATIC ITEMS
1. What are the joint cognitive system issues in question?	Pitfalls: superficial labels
2. Explicitly map target and test situation relationship	Problem design, stimulus sampling
	Pitfalls: psychologists fallacy, cover stories
3. Shape the test situation to address the general JCS issues	Scenario design
	Artifacts as tool for discovery
	Procedure to externalize what underlies adapted activities
4. Map canonical behavior; how does one prepare to be surprised?	
5. Collect the data (update 4 as needed)	
6. Collate multiple raw data sources to construct base protocol	Analyze each participant/team's process
	Pitfalls: get lost in the raw data
7. Use concepts from 1 to build formal protocol	Pitfalls: get lost in the language of the domain; excessive micro-encodings; hindsight bias
	Distinguishing interpretations from observations
8. Cross-protocol contrasts	Aggregate across teams and problems
	Role of cognitive simulation, neutral observers
9. Integrate with other converging studies to build a functional model	

The processes in Figure 6 capture the heart of Neisser's perceptual cycle as a model of robust conceptualization and revision (see also Figure 1): moving up abstracts particulars into patterns; moving down puts abstract concepts into empirical jeopardy as explanatory or anticipatory tools in specific situations. This view of re-conceptualization points out that what is critical is not one or the other of these processes; rather, the value comes from engaging in both in parallel. When focused on abstract patterns, shift and consider how the abstract plays out in varying particular situations; when focused on the particular, shift and consider how the particular instantiates more abstract patterns (see also the discussion on pp. 2-3). This is a basic heuristic for functional synthesis, and the interplay of moving up and down helps balance the trade-off between the risk of being trapped in the details of specific situations, people, and events, and the risk of being trapped in dependence on a set of concepts which can prove incomplete or wrong (see also the discussion of revising assessments in anomaly response pp. 74-75).

While the ability to stage problems is fundamental, a powerful resource for functional analysis/synthesis is to observe the process of absorbing new artifacts into practice. Similarly, one can observe how changes in the distributed system, such as linking a new a collaborator at a distance through a technology portal, affects the coordination demands and capabilities. These points of change may occur naturally, or the researcher may introduce points of change in how they provide new artifacts or shift the relationships between agents for a staged problem (e.g., Layton et al., 1994; Cook & Woods, 1996). Focusing observation on these points of change can reveal a great deal about the nature of practice.

When investigators probe the nature of practice by changing the kinds of artifacts practitioners use to carry out their activities, the intent is not to evaluate the artifact as a partially refined final product, but rather to use it as a tool for discovery, i.e., as a wedge to break through the fluency of current adaptations in order to see the basic demands of practice and factors in JCSs.

Artifact-based methods have great potential because introducing new technology into fields of ongoing activity always serves as a kind of natural experimental intervention (Flores et al., 1988; Carroll et al., 1991). Prototypes of new systems function as a kind of experimental probe to better understand the nature of practice and to help discover what would be useful (even though their parallel status as partially refined final products usually dominates the perspective of evaluators; Woods, 1998). Artifact-based methods for studying JCSs depend on a model of how the artifact potentially functions as an affordance for work, or how the change in artifacts changes the work of the JCS (e.g., see Watts-Perotti and Woods, 1999; Guerlain et al., 1999). The artifact functions as a hypothesis of what would prove useful (Woods, 1998). It is important to re-represent the research base in ways that identify provisional design directions and to define criteria for assessing the promisingness of that direction for supporting the demands of work in the future (Roesler, Woods & Feil, 2005).

In the final analysis, Staged World studies are fundamentally observational and discovery oriented—letting the world tell us how it works rather than playing 20 questions with nature in an artificial lab (Newell, 1973; Flach et al., 2006). Building on top of the fundamental observational nature of this family of methods, investigators can engineer contrasts through sets of scenarios, changes in artifacts, and changes in collaborative links—again, if they can draw on knowledge about the relevant problem space, affordances, and functions of JCSs at work underlying these changes. But it is important not to be confused at this point into thinking that these contrasts convert a Staged World observational study into a Spartan lab experiment. Each of the things contrasted (problem demand, affordance, collaborative function) is still a complex conglomerate in its own right with many degrees of freedom. Avoid the error of mistaking complex factors for primitive elements which require no further analysis.

While these three families of Natural History, Staged Worlds and Spartan Lab methods differ in many ways, they are all grounded on the primacy of observation. The most basic role of the scientist is that of trained observer—one who sees with a fresh view, who wonders, child-like—*why is it this way, how does it work?*—

making fresh connections and generating new insights. Similarly, all three classes require meeting basic criteria (Salomon, 1991):

• Means to establish warrant—How do you know? Why should one accept your findings, observations, conclusions, and interpretations?
• Standards of quality—How do you discern a competent from an incompetent investigation given the uncertainties of research and modeling?
• Means to facilitate generalizability—What is this a case of? What other cases/situations does this apply to?

However, each family represents a different way to balance the basic tradeoffs across uncertainties and authenticity (Figure 5) so that each family meets the above criteria in very different ways (Hoffman & Woods, 2000).

Also basic to all three families is the mapping between the target situation to be understood and the test situation where observations are made. In Natural History techniques, the target is examined relatively directly, though subject to a variety of choices about who observes what, when, and from where. What is particularly interesting is that all of the constraints and issues that apply to Natural History methods apply to Staged World methods, plus more. And all of the constraints and issues that apply to Staged World methods flow to Spartan Lab studies, plus more. Ironically, moving from *in situ* to scaled worlds to artificial worlds brings sources of additional leverage, but at the cost of new dependencies related to authenticity in mapping between the target situations and test situations.

CONVERGING OPERATIONS

Cognitive Task Analysis is more than the application of any single technique for Cognitive Task Analysis.
Potter et al., 2000

Different studies within and across the three families of methods in Figure 5 become part of a larger series of converging operations that overcome or mitigate the limits of individual studies. This is based both on using different techniques to examine a single field of practice and on examining the same theme across different natural laboratories. As a result, we speak of studying work in context as a bootstrapping process that uses a set of converging operations. The bootstrap means that each technique used provides tentative knowledge that facilitates or expands our ability to use another technique to wedge open how the JCS functions (Potter et al., 2000). For example, elicitation techniques that ask practitioners to share stories of expertise in action (Klein, 1999) are often used to help identify what makes situations difficult. This information is important for designing problems embodied in scenarios for Scaled World simulation or for walkthrough elicitations with other practitioners.

Richard Cook and colleagues' studies of infusion devices in health care provide an example of converging operations. The line of work began with a series of critical incidents—building a corpus of cases as a natural history technique (Cook et al., 1992; Cook et al., 1998). Parallel studies of physician interaction with these devices, drawing on concepts such as poor observability, limited directability, and the danger of getting lost in virtual data spaces, illustrated poor design for coordination and resilience and classic strategies for coping with complexity when artifacts are clumsy (Moll van Charante et al., 1993; Nunnally et al., 2004; Nemeth et al., 2005). The engine that drove the studies was the going back and forth between observing how physicians handled various *staged* situations and the infusion devices and observations of device use *in situ*.

The work of Nadine Sarter on human-centered automation is a complete example of the combination of Natural History and Staged World techniques— building a corpus of cases, designing problems, discovery of concepts about how JCSs work in general, and defining promising new design directions (including data that provide strong warrant that these design directions are promising practically and scientifically). The natural laboratory was the *flight deck* where *pilots' interact with cockpit automation.* The program began with Natural History techniques—building a corpus of cases (e.g., Sarter & Woods, 1997), jump started by previous research results on making automation a team player (e.g., Norman, 1990). These results provided a basis for Staged World studies (e.g., Sarter & Woods, 2000) that provided repeated observations of how line pilots handled challenges to coordination with their automated partners. The results identified specific problems that needed solutions—the problem of indirect mode transitions and mode awareness (Sarter & Woods, 1995)—and provided new concepts about JCSs in terms of the factors that lead to *automation surprises* (Sarter et al., 1997; see Chapter 10).

These results, with converging results from studies of human-automation coordination breakdowns in other settings, identified design directions that showed promise for illustrating how to make automated systems team players—how to increase observability of what the automation will do next, and expand the directability of the automated systems as resources to achieve operational goals (e.g., Woods & Christoffersen, 2002). In particular, Sarter developed and explored the promise of tactile cues to solve the problem of mode awareness on flight decks (Sklar & Sarter, 1999; Nikolic & Sarter, 2001). These results indicate, in general, how to build JCSs that exhibit effective control of attention in any multi-threaded situation where people work with other machine and human agents (Sarter, 2002), which is a basic affordance in JCSs (Woods, 2005c).

THE PSYCHOLOGIST'S FALLACY

Because technical work ... is so poorly understood, policy makers routinely fall back on stereotypes or images of work ... in order to make sense of technical work. The potential cost of misunderstanding 'technical work' is the risk of setting policies whose actual effects are 'not only unintended but sometimes so skewed that they exacerbate the problems they seek to resolve. ... Efforts to reduce 'error' misfire when they are predicated on a fundamental misunderstanding of the primary

> sources of failures in the field of practice [systemic vulnerabilities] and on *misconceptions of what practitioners actually do.*
>
> Barley & Orr, 1997, p. 18, *emphasis added*

As one orchestrates a series of investigations, as in the program of research developed by Sarter, there is a variety of pitfalls that must be evaded. The most basic pitfall is what William James (1890), over 100 years ago, called the Psychologist's Fallacy. Updated to today, this fallacy occurs when well-intentioned observers think that their distant view of the workplace captures the actual experience of those who perform work in context. Distant views can miss important aspects of the actual work situation and thus can miss critical factors that determine human performance in that field of practice. Integrating Natural History techniques into a program of research is a strong guard against the Psychologist's Fallacy.

There is a corollary to the Psychologist's Fallacy that is another pitfall in pursuit of authenticity.

> … Avoid assuming that those we study are less rational or have a weaker grasp on reality than we ourselves. This rule of method, then, asks us to take seriously the beliefs, projects, and resources of those whom we wish to understand.
>
> Law & Callon, 1995, p. 281

Managers, designers, researchers, and regulators often feel they are immune to the processes they study and manage, especially when the triggers for investing in studies are human "error," "limits," and "biases." This fallacy blocks effective functional analysis as it blinds the observer from seeing how behavior is adapted in a universe of multiple pressures, uncertainty, and finite resources (Woods et al., 1994). Learning about JCSs at work begins with adopting the point of view of practitioners in the situation before outcome is known (Dekker, 2002).

The search for authenticity reveals a trade-off: the stronger the shaping of the conditions of observation, the greater the disruption of the thing observed and the greater the risk of Psychologist's Fallacy and its analogs. Natural History methods, then, are not "weak" as in "poor substitutes" for the real methods of science or "necessary evils due to practical limitations." Rather, because they are weak in shaping the conditions of observation, Natural History methods are a fundamental contributor to the goal of transcending limits to authenticity in order to capture how the strategies and behavior of people are adapted to the constraints and demands of fields of practice.

Because Natural History methods are weak in shaping the conditions of observation, they need to be coordinated with investigations that more strongly shape the conditions of observation based on tentative, partial knowledge of the essential factors in JCSs at work. But Staged World techniques are weak in the sense that they depend on strong commitments to models of how cognition adapts to artifacts and demands of work. And Spartan labs are even weaker in the sense that they presume a great deal of knowledge about what can be safely left out of the greatly reduced experimenter-created situation. Thus, converging operations that coordinate multiple techniques and balance different ways to shape the conditions of observation serve as guards against the vulnerabilities captured in James' Psychologist's Fallacy.

Chapter 6

Functional Syntheses, Laws, and Design

PROPERTIES OF FUNCTIONAL SYNTHESES

Functional synthesis is a mode for research that is critical to understanding systems that are only partially decomposable, dynamic, and adaptive, such as joint cognitive systems (JCSs). Functional synthesis is empirical since it is based on a variety of observations collected in different ways. It is theoretical as it proposes a model of how something functions in order to achieve goals.

The functional accounts in "Being *Bumpable*" and in Hutchins (1995b) and Patterson et al., (1999) illustrate (and should be judged against) the following characteristics of functional models captured in Table 3.

Functional syntheses generate models based on patterns that are abstracted from observation. These candidate accounts can be subjected to a critical examination through argument based on (a) re-interpretation of what would account for patterns and (b) abstracting additional patterns. Functional syntheses motivate and guide empirical confrontations with the field of practice. The functional model serves to abstract the particulars of a setting in a way that supports bounded generalities. Functions serve as a tentative condensation of what has been learned (subject to critical examination and revision) based on a set of empirical observations and investigations.

Functional modeling also has the advantage that it can be used to assess the degree of coupling in JCSs—the more intertwined the relationships between structure and function, the more complex the system operationally (and the less the system is decomposable into almost independent parts).

Functional syntheses provide models of how systems are adapted; hence the use of the word—"well"-adapted—in the statement of the Law of Fluency is deliberately meant to provoke reflection ("well"-adapted work occurs with a facility that belies the difficulty of the demands resolved and the dilemmas balanced). Behavior of JCSs is adapted to some purposes, potential variations, and constraints in the world of work (this is also a facet of the meaning of another law that governs JCSs at work—there is no such thing as a cognitive vacuum—the work system is adapted to some purposes, variations, constraints, even if they are not the ones that outsiders see or desire; p. 11 this volume). Despite being "well"-adapted, the current forms of adaptation may be incomplete, misbalanced, or brittle, relative to alternative forms of adaptation. Systems can be better or worse adapted to a set of constraints. Adapted systems can be subject to variations outside the planned-for or designed-for ranges and kinds of disturbances. Systems can be brittle or robust when near the margins or boundaries of adaptation. Systems can be aware of or blind to the limits on their adaptive range. Thus, functional accounts do not search for optimal or best strategies, but rather help generate moves or project the consequences of moves that would shift or increase the adaptive powers of the JCS (and, thus, are needed to steer change beyond mere trial and error and beyond chasing the tail of technology change (Woods & Tinapple, 1999).

Table 3. Characteristics of Functional Syntheses of Work (Woods, 2003)

1. A functional synthesis is context bound, but generic.
2. A functional synthesis is tentative; it can be overturned or the patterns re-interpreted. It generates candidate explanations that attempt to meet one kind of sufficiency criteria—an ability to carry out the functions in question.
3. A functional synthesis is goal oriented; purposes serve as a frame of reference or point of view; the functions of pieces change with purpose (i.e., purposes provide varying perspectives).
4. Since functional syntheses include purpose, they provide accounts of how multiple goals and potential conflicts across goals present tradeoffs and dilemmas to practice.
5. Functional syntheses concern dynamic processes so that change is fundamental. Such accounts lend themselves to representation as general storylines or story archetypes.
6. Functional syntheses are inherently multi-level; phenomena at one level of analysis are situated in a higher level context and are present because of what they contribute in that context.
7. Functional syntheses involve the mutual interaction and adaptation of agent and environment (ecological).
8. Functional syntheses, since they are models of adaptation, support projections of how systems will respond when changes are introduced so that side effects of change can be identified without waiting for consequences to occur.
9. Functional syntheses reveal vulnerabilities for under- versus over-adaptation.
10. Functional syntheses do not imply "correct," or "ideal," or "fixed" (i.e., that "it has to be that way" or that "it is best done that way"). Rather, they capture what behavior is adapted to (variations and constraints) and how it is adapted to handle these constraints. Understanding how the system has adapted allows outside and broader perspectives to consider how to expand the adaptive power of the system. Thus, functional syntheses are valued in the their ability to generate guides for constructive change.

ON LAWS THAT GOVERN
JOINT COGNITIVE SYSTEMS AT WORK

Functional syntheses capture generic patterns. But sometimes, as we introduce and refer to different cases of JCSs at work throughout this book, we have referred to some patterns as if they can be considered "Laws that Govern JCSs at Work" (Woods, 2002). We use the label "law" in the sense of a compact form of generalization of patterns abstracted from diverse observation. Patterns that can be tentatively labeled "laws" provide basic first principles that inform and guide the study and design of JCSs at work. Extracting, sharing, using, and questioning such first principles is critical if we are to escape being trapped in the specifics of particular technologies, particular operational practices, particular work settings.

Also note that the laws we introduce are control laws (Hollnagel, 1993; Flach et al., 2004; *JCS-Foundations*, Chapter 7). Remember, *control* or *regulation* refers to steering in the face of changing disturbances (as in *cybernetics* from the Greek for "steersman;" Wiener, 1948). Control laws have an odd quality—they appear optional. Designers of systems that perform work do not have to follow them. In fact, we notice these laws through the consequences that have followed repeatedly when design breaks them in episodes of technology change. The statements are law-like in that they capture regularities of control and adaptation of JCSs at work, and they determine the dynamic response, resilience, stability or instability of the distributed cognitive system in question. While developers may find following the laws optional, the consequences that follow predictably from breaking these laws are not optional. Many times these consequences block achievement of the performance goals that developers and practitioners, technologists, and stakeholders had set for that episode of investment and change.

Such laws guide how to look and what to look for when we encounter JCSs at work. They "jump-start" studies of new settings. The first principles allow new studies of settings to set more ambitious targets for observation by building on past generalizations. What is observed then may question, shift, expand, or conflict with past generalizations; this process of forming, questioning and revising basic patterns is fundamental to healthy areas of inquiry in science.

CHALLENGES TO INFORM DESIGN

Ultimately, the purpose of studying JCSs is to inform design—to help in the search for what is promising or what would be useful in changing fields of practice (Woods, 1998; Roesler et al., 2005). Thus, methods for studying JCSs at work are means to stimulate innovation and not only ways to build up the pattern base (Figure 1; Woods, 2002). In many ways, the impact of design on studies of how JCSs work is a reminder that in cyclical processes like research (Figure 1), the cycle can rotate in two directions. What you want to do next (influence design and innovation) rotates back to influence how you study and model a JCS, especially when previously designed artifacts are part of the JCS under study and are undergoing change.

Studies of JCSs usually occur as part of a process of organizational and technological change spurred by the promise of new capabilities, dread of some paths to failure, and continuing pressure for higher levels of performance and greater productivity (systems under "faster, better, cheaper" pressure). This means investigators, through observing and modeling practice, are participants in processes of change in those fields of practice.

To become integral participants in generating new possibilities for the future of a JCS, the families of methods must be extended to meet the challenges of design. There are five challenges design imposes on how to study JCSs at work. When considering a program of converging operations across varying methods to understand a JCS at work, the set is valuable to the degree that it helps meet these challenges from design:

• The *leverage* problem—How do studies of JCS at work help decide where to spend limited resources in order to have significant impact (since all development processes are resource limited)?

• The *innovation* problem—How do studies of JCS at work support the innovation process (the studies are necessary but not sufficient as a spark for innovation)?

• The *envisioned world* problem—How do the results that characterize cognitive and cooperative activities in the current field of practice inform or apply to the design process since the introduction of new technology will transform the nature of practice? (A kind of moving target difficulty.)

• The *adaptation through use* problem—How does one predict and shape the process of transformation and adaptation that follows technological change?

• The problem of *"error" in design*—Designers' hypotheses, as expressed in artifacts, often fall prey to William James' Psychologist's Fallacy, which is the fallacy of substituting the designer's vision of what the impact of the new technology on cognition and collaboration might be, for empirically based but generalizable findings about the actual effects from the point of view of people working in fields of practice (Woods & Dekker, 2000).

To summarize: Discovering Patterns (Chapter 4, 5 and 6)

The hallmark of any science is empirical confrontation – one observes to note patterns; one generates ideas to account for the variability on the surface; and one subjects these ideas to empirical jeopardy. Different traditions in method balance these trade-offs differently, but the ultimate test is one of productivity – do the methods help generate patterns and ideas about what are the essential variables and processes that account for the breadth and diversity on the surface?

CSE is discovery-oriented and open to the recognition of patterns and the discovery of tentative abstractions that underlie the surface variability. The puzzle of discovery is captured in the phrase, "how to be prepared to be surprised." Methods to study JCSs at work ultimately function to help observers prepare to be surprised. Observation is jump started by previous abstractions: What is this setting or activity an example of? How does it fit or differ from patterns in the research base? For abstraction of patterns to occur, field settings function in parallel as natural laboratories. Methods vary in how one shapes the conditions of observation across a series of converging operations that overcome or mitigate the limits of individual studies. This is based on using different techniques to examine a single field of practice and on examining the same theme across different natural laboratories.

Through these techniques, to study JCSs at work is to build a functional synthesis that captures how the behavior and strategies of practitioners are adapted to the various purposes and constraints of the field of activity. In the final analysis, insight is precious, and the techniques for studying JCSs at work constitute a program for facilitating insight.

Additional resources: Readers should examine in great detail two other cognitive task syntheses and the two process tracing Staged World studies that illustrate the central concepts presented here. See Hutchins (1995b) & Patterson et al. (1999); then examine Roth et al. (1987) & Johnson et al. (2001).

A living multi-media resource set on the concepts for how to study JCSs at work can be found at http://csel.eng.ohio-state.edu/woodscta. Studying Cognitive Work in Context: Facilitating Insight at the Intersection of People, Technology and Work, by D. D. Woods. Produced by Woods, D., Tinapple, D., Roesler, A. & Feil, M. (2002).

Patterns in How Joint Cognitive Systems Work

CSE is committed to abstracting general patterns across specific settings and various episodes of change. The following chapters focus on basic concepts that have emerged from over two decades of CSE research. In each case, studies challenged conventional beliefs and provided a new understanding of how joint cognitive systems adapt to cope with complexity.

Chapter 7

Archetypical Stories of Joint Cognitive Systems at Work

One of the great values of science is that, during the process of discovery, conventional beliefs are questioned by putting them in empirical jeopardy. When scientists formulate ideas and look at the world anew through these conceptual looking glasses, the results often startle us. The new views are particularly valuable when they stimulate and steer the processes of innovation in worlds of work.

CSE arose in just this way—stimulated by surprising reverberations of vectors of change to provide new sets of conceptual looking glasses. Looking at changing worlds of work through the core concept of the JCS as core unit for analyses of coping with complexity—that provoked insights about the nature of work and stimulated innovations about promising design seeds.

In studying a workplace, we are out to learn how the more or less visible activities of practitioners are parts of larger processes of collaboration and coordination, how they are shaped by the artifacts and in turn shape how those artifacts function in the workplace, and how their activities are adapted to the multiple goals and constraints of the organizational context and the work domain. When we study examples of JCSs at work, given the basic values of CSE and given the basic methods for discovering how JCSs work, we begin to notice recurring concepts central to the descriptions of general patterns that follow.

DEMANDS AND ADAPTATION

One word used frequently in describing JCSs at work is *demands*. Demands are general aspects of situations that make goal achievement difficult. Demands are seen in the difficulties that arise as situations unfold to challenge the performance of any JCS.

Demands are a kind of constraint on how to handle problems. In turn, skills—routinization and resilience—develop through and around constraints. Discovering how a JCS works involves the process of connecting practitioner behavior to constraints as a description of processes of change, adaptation, and equilibrium. In other words, understanding adaptation depends on understanding the demands adapted to; to see the demands in classes of situations, one must observe processes of adaptation across varying but related contexts. This co-dependency is traced out, for example, in the cognitive task synthesis of "Being *Bumpable*" in Chapter 3.

Demands are concepts about the nature of work—for example, multi-threaded situations present demands on control of attention, but they play out in specific situations in particular fields of practice. Thus, the cross-level connections are important. One can only see the general patterns defined by demands through how they play out in concrete situations; the concrete situations appear infinitely variable and unique, yet underneath they spring from and express a small number of basic patterns (as Alexander captured in articulating a pattern-approach toward a generalizable and design-oriented research base). This means that abstracting demands is necessary for a formal, concept-based language of description for how situations unfold (Figure 6).

A problem space, then, is not simply the states of the world as in a game (the board configurations in chess); rather, a problem space is a model of the way demands are inter-connected and how they play out in particular situations in a specific field of practice. In particular, such a description captures the tradeoffs and dilemmas inherent in practice and how these may be balanced or mis-balanced in specific situations (Woods et al., 1994).

Demands are uncovered (they are not directly visible in practitioner activity) by tracing processes of change and adaptation and settling into equilibria at different scales. Processes of change disrupt balances; as a result, change triggers adaptive responses by different stakeholders as they work around complexities and as they attempt to exploit new capabilities. Eventually, a new equilibrium is established. To abstract demands from particular situations, one looks for story lines within these cycles of adaptation. Throughout this book, general story lines such as over-simplifications, reframing, automation surprises, and others appear as parts of larger cycles of adaptation (e.g., the Law of Stretched Systems).

By starting to build up descriptions of demands, one can begin to trace out the three relational properties of a JCS at work: *affordance, coordination*, and *resilience*. Each of these properties expresses a relationship between people (and other agents), artifacts (technological capabilities), and specific work settings (domains). This book captures samples of patterns (or story archetypes) in coordination, or its contrast, miscoordination, samples of patterns in resilience, or its contrast, brittleness, samples of patterns in affordance, or its contrast, clumsiness. While a story may start out emphasizing one of these three themes, in the end each story of a JCS at work expresses aspects of all three.

Affordances

An affordance is a relationship between observers or actors and the artifacts used to meet the demands of work. The best way to define affordance is to examine specific cases, positive and negative, such as the story of a serendipitous affordance of an auditory display of rod position in a nuclear power plant control room (see pp. 15-16). In that example, the artifact is the physical device and its behaviors—in this case, audible clicks from mechanical counters that indicate rod position. The affordance arises in how the device is used to handle the demands expressed in anomaly response situations. First, the behavior of the device is connected to a class of changes in the monitored process and, importantly, provides feedback on how an automatic system is handling changes in the monitored process. Second, the observer can pick up the change in the monitored process and in the control of the monitored process by another agent without disrupting other lines of reasoning or work. In this way the artifact supports two critical basic and generic affordances for JCSs at work: observability—feedback that provides insight into a process, and directed attention—the ability to re-orient focus in a changing world (see Chapter 12 for more on generic support requirements or general affordances for JCSs at work).

As seen in this example, affordances are about the **fit** across the triad of demands, agents, and artifacts. Hence, when an artifact provides an affordance we generally describe a *direct mapping* or direct correspondence of artifact and demands for a role. When an artifact is poorly designed (results in a poor fit), we generally describe how tasks must be accomplished through an indirect series of steps. The result of such indirect steps is a more effortful, deliberative, or "clumsy" coupling across the triad (e.g., Woods et al., 1994, chapter 5).

Affordances are not properties of artifacts per se; affordances are relationships between an observer/actor and the artifact as it supports the practitioner's activities and goals in context. In addition, the example of the auditory display for rod control/nuclear control room illustrates another property of affordances. Affordances cannot be completely specified through design intent and activity; ultimately, the practitioner participates in and creates the affordance through how they wield the artifact to establish the affordance. This is often expressed in one or another variant on the phrase: *artifacts become tools through use*. This example of an affordance is an extreme one because the value of the auditory display was completely unintended in design (for that matter, even the idea that an auditory display had been designed was unintended in this case).

The concept of an affordance has proved hard to define concisely since it was first introduced and coined by Gibson (see Gibson, 1979). Why? Because to say that an object has an affordance is to claim a functional model that relates artifact to practitioner to demands given goals (see pp. 55-56 for a discussion of what constitutes a complete description of, criteria for, and means for debating functional models). Developers almost always describe intended affordances for the artifacts they are working to realize. Often, these stated or implicit claims for affordance are quite oversimplified with respect to the actual dynamics of JCSs at work (Woods & Dekker, 2000). But this problem of weak or poor functional models does not result only from poor JCS design processes; universally, all functional claims during design can only be measured and tested relative to artifacts in use to meet demands.

This orientation to future behavior and testing through future impact subjects the design of affordances to the envisioned world problem—claims for designs are stories about future operations. To cope with this constraint on design, claims to affordance are based on evidence from functional syntheses derived from studies of past cycles of adaptation, and are projected ahead based on anticipation of how practitioners and managers adapt to exploit capabilities or workaround complexities in use—design tells, and shares, stories of the future (Roesler et al., 2001; Murray, 2005).

Coordination

Coordination and its related labels, collaboration and cooperation, are similarly complex and difficult to define concisely. Coordination points to the basic finding that work always occurs in the context of multiple parties and interests as moments of private cognition punctuate flows of interaction and coordination (Hutchins, 1995a). The idea that cognition is fundamentally social and interactive, and not private, radically shifts the bases for analyzing and designing JCSs that work, as well as for reconsidering the relationship between people and computers in work. The concept of a JCS originated as a call to study and design how work is distributed and synchronized over multiple agents and artifacts in pace with changing situations.

Note how the affordance of the serendipitous auditory display in the old nuclear control center (pp. 15-16) is also about coordination and synchronization as situations evolve. The auditory display provided information about how another agent was working to handle variation in the monitored process—making another agent's activities observable. The information picked up was valuable because it helped to specify when the practitioners needed to re-direct attention to new tasks or priorities given the flow of other competing tasks and priorities—a kind of synchronization. Also note another aspect of this case that reflects a general characteristic of coordinated activity: the smoothness of flow (despite the huge range of varying details) in the situation described—there are no explicit conversations; no sequential and explicit analysis of data to assess the situation; no effortful decision to investigate (cf. Law of Fluency).

Synchronization and the smooth flow of highly coordinated activities are particularly difficult to study, model and measure because by definition they are highly sensitive to temporal details of the specific situation at hand (De Keyser, 1990; Nyssen & Javaux, 1996; Carroll et al., 2003; Ho et al., 2004). This difficulty is one of the reasons why we have to remind designers of distributed work systems that "coordination costs continually." This "law" captures how high levels of coordinated activity are highly contingent on the temporal flow of events and activities and therefore require ongoing practice in order to maintain high proficiency.

The recognition that work is fundamentally distributed challenges conventional beliefs about the relationship between people and automation (the substitution myth and the repeated failure of function allocation based on Fitts' list accounts of what people do well versus what machines do well; see *JCS-Foundations*, pp. 101-102; 121-122). This is captured in one of the base slogans of CSE that emerged around 1990 (Malin et al., 1991; Roth, Malin, & Schreckenghost, 1997)—*how to make automated systems team players*. The difficulty is that algorithms are designed as autonomous performers; hence, their participation in the basic functions as a competent team player—grounding, relevance, intentions, responsibility, and others—is limited (Woods, 1996; Woods & Sarter, 2000; Klein et al., 2004). These basic constraints on automata have led to the recognition that there is a fundamental asymmetry in the competencies for coordinated activity between people and machines. Machine agents cannot be counted on to participate fully in the functions of coordinating teams in the same sense that other people can—coordinative functions such as common ground, inter-predictability, synchronization, and others (Brennan, 1998; Woods et al., 2004; Klein, Feltovich et al., 2005).

Resilience

The third core concept about how JCSs work is resilience (see Hollnagel, 1998; Hollnagel et al., 2006). Given that variability is the key constraint that leads to models of work as forms of control or adaptability (*JCS-Foundations*, chapter 7), resilience relates to the idea of unanticipated variability or perturbations (Woods et al., 1990; Woods, 2006a). Systems adapt in order to be increasingly competent at handling designed-for-uncertainties. This results in a "textbook" performance envelope that consists of how systems have adapted formally in order to be competent at handling designed-for-uncertainties. Resilience is concerned with how a system recognizes and handles situations which challenge or fall outside that area of textbook competence (Woods 2006a). Such inevitable episodes of unanticipated variability or perturbations call into question the model of competence and demand a shift of processes, strategies and coordination. Such episodes are inevitable because of change and because routines are inherently underspecified (Suchman, 1987; Woods et al., 1994).

This aspect of resilience is fundamental to all adaptive systems, at any level of analysis, because all systems are under some time varying pressure to meet new resource pressures and performance demands (Law of Stretched Systems and the Bounded Rationality Syllogism; see pp. 18 and 2, respectively). The simultaneous demands for efficiency and resilience create a fundamental trade-off or dilemma:

• All adaptive systems have means to improve routines in order to better handle frequently encountered situations.
• All adaptive systems have means to monitor or check that the routine deployed fits the actual situation at hand; otherwise the risk of the error of the third kind is too high (solving the wrong problem; Mitroff, 1974; cf. also the discussion of literal-minded agents in Chapter 11).

These two demands compete to some degree, however, so that all adaptive systems try to adjust to find a balance across these two processes. CSE studies of breakdowns in JCSs quickly recognized that there are two basic ways this balance fails (Woods et al., 1987; Woods & Shattuck, 2000): *under-adaptation or over-adaptation*. In cases of under-adaptation, routines and plans continued on even when the situation had changed or new evidence indicated the need to revise assessments and responses. In cases of over-adaptation, the breakdown took the form of ad hoc adaptations that failed to take into account the larger perspective, failed to address broader goals or intent of the plan behind the routines, and failed to balance constraints across larger spans of control and fields of view. For example, over-adaptation often includes missing side effects of changes when modifying plans in progress. Thus, resilience is the ability to anticipate and adapt to the potential for surprise, given the inherent limits in the model of competence that drives routinization in the face of complexity and change.

This idea of a balance of routinization and resilience is actually a quite old one in models of human performance (and now extended to JCSs). For example, Jens Rasmussen (1979) captured this notion at the beginning of CSE in his S-R-K framework, which described three parallel and simultaneous levels of control in work (skill, rule, knowledge). Skills and rules referred to two types of routinization that adjusted the competence envelope. The third level (unfortunately, labeled *knowledge*) referred to checks on the potential for brittleness by monitoring for whether the routines being deployed actually fit the situation at hand given change. These processes looked for emerging evidence about the limits to the comprehensiveness of the model of competence as encoded in the skills, procedures, and plans in progress and provided a means to adapt when disruptions to plans occurred. The key in the model is the general need to balance multi-level processes to avoid under- or over-adaptation.

Story Archetypes in 'Being *Bumpable*'

The trio of core concepts—affordance, coordination, resilience—exist only as relationships at the intersection of people, technology and work. To solidify your grasp of these central ideas, re-visit a strong functional synthesis of a JCS you have studied and examine how the account is built on exploring how these three concepts play out in that setting. For example, revisit "Being *Bumpable*" in Chapter 3. This synthesis of how a joint cognitive system functions reveals each process at work in the setting of interest:

• Resilience—the brittleness apparent in the incident helps to highlight the normal sources of resilience, and highlights as significant the adaptations and work that even practitioners themselves may not reflect on.

• Coordination—the flow of information across distributed parties even with a central actor; the multiple roles played by a single agent at center stage in the story; the synchronization demands in terms of timing across agents and tasks.

• Affordances appear in the role of the *Bed Book*, but more generally the account allows one to consider what might be promising or to know how to test prototype new artifacts as they instantiate hypotheses about what might be useful. For example, new concepts might support anticipation to help manage the multiple threads of work or new concepts might support avoiding or handling the bottlenecks that can emerge given that the *bedmeister* has multiple roles and responsibilities.

By exposing processes of coordination, resilience and affordance (or their opposites), a functional synthesis of JCS highlights basic demands in work such as the handling and switching among multiple changing threads; the need for anticipation to synchronize across agents and adapt to change; the need to shift priorities across tasks in time; and vulnerabilities to overload, relying on stale information, bottlenecks, and more. Results in the ICU as one natural laboratory can then be compared, contrasted and synthesized across studies of the same themes in different settings and through different conditions of observation to create a picture of what contributes to competence and expertise on different themes about JCSs (the top layers in Figure 6).

Notice that to pry open the interdependencies across a JCS at work, one shifts across the three factors in the search for wedges to pry open and discover how the JCS works. For example, "Being *Bumpable*" traces a coordination breakdown and deconstructs a term of practice in order to reveal the processes of resilience. The Hutchins' (1995b) study of "How a Cockpit Remembers Its Speed" began with a task and considered how it was distributed between individuals, crew and external representations. The cognitive task synthesis revealed the power of some apparently small and unremarkable devices (speed bugs around a round analog meter) to help the JCS keep track of changing speeds during a descent (synchronization). The directness of the fit or mapping stands out to illustrate an affordance in contrast with the clumsiness of artifacts designed to make up for human memory limits (and highlighting the contrast between Hutchins' functional account and alternative accounts based on human memory). The Patterson et al. (1999) study of voice loops in space shuttle mission control had the purpose to understand what an artifact (voice loops) afforded practitioners in anomaly response based on the strong suspicion that this artifact's role in the effective performance was obscured by newer and more faddish technologies. The study of affordances proved to be an account of how coordination across groups was central to meet the demands inherent in anomaly response (resilience).

Understanding how the JCS is adapted (a functional synthesis) is powerful because it allows us to escape the limits of incremental analysis and design (Woods & Tinapple, 1999). Systems development is often trapped into just describing the activities required for carrying out the tasks of today's systems or trapped into making the next technology development fieldable into ongoing operational settings or trapped into incremental design by rapid trial and correction (made more attractive by technologies for rapid prototyping). Functional synthesis of JCSs at work is an innovation strategy needed to meet the five challenges to inform design (p. 58).

To consider more about patterns in JCSs at work, we will begin with some of the demands of one class of situations—anomaly response (Malin et al., 1991).

To summarize: Archetypical Patterns

Building a useful research base for design can be seen as a process of abstracting general, recurring story lines about how systems at work are resilient or brittle, how sets of activities and agents are coordinated or fragmented, how artifacts trigger adaptive responses as they provide either affordances to meet the demands of work or new complexities to be worked around. Finding, seeing, abstracting, and sharing these kinds of story archetypes put the four key values of CSE into practice (Chapter 1): authenticity, abstraction, innovation, and participation.

Additional resources: Readers should examine the preface to Christopher Alexander et al.'s "A Pattern Language" (1977) on how patterns serve as a means to connect research and design. To capture, share and use abstract patterns leads one to consider storytelling as a mode of representation to support collaboration across different stakeholders in R&D. Storytelling is central to innovation as it seeks to re-conceptualize what it means to practice as new technological powers enable or drive change. See Roesler et al., 2001 for one treatment of how narrative processes and representations can support collaborative envisioning in R&D at http://csel.eng.ohio-state.edu/animock. For an example, Cook et al. (1998) organize approaches to improve patient safety as a contrast between different stories given hindsight bias.

There are many sources available to examine processes in coordinated activity and cooperative work (e.g., Olson et al., 2001). To further examine the concept of affordance it is best to carefully consider the original source—Gibson (Gibson, 1979 provides a comprehensive account of his ideas). Hollnagel et al. (2006) provides the first broad treatment of resilience in work systems (focused on safety issues).

Chapter 8

Anomaly Response

Cascading Effects in an Episode of Anomaly Response

Several minutes into the ascent phase of a space shuttle mission (Figure 7), one of the flight controllers responsible for monitoring the health and safety of the mechanical systems in mission control centre noticed an anomaly – an unexpected drop in hydraulic fluid in an auxiliary power unit (APU). The personnel monitoring immediately recognized that the anomalous reading pointed to a hydraulic leak. Did this anomaly require an immediate abort of the ascent? In other words, she needed to assess—How bad was the leak? Was it a threat to the safety of the mission? What were the relevant criteria (and who knew them, and where did they reside)? The mechanical systems controllers did a quick calculation that indicated the leak rate was below the predetermined abort limit – the mission could proceed to orbit. The analysis of the event relative to an abort decision occurred very quickly, in part because the nature of the disturbance was clear and because of the serious potential consequences to the safety of the astronauts of any anomaly at this stage of a shuttle flight.

As the ascent continued, a second collection of demands and activities was intertwined and went on in parallel. The controllers for the affected system informed the Flight Director and the other members of the mission control team of the existence of the hydraulic leak and its severity. Because of the nature of the artifacts for supporting coordination across controllers and teams (voice loops, see Patterson et al., 1999), this occurred quickly and with little overhead costs in work and attention. The teams also had to consider how to respond to the anomaly before the transition from the ascent to the orbit phase was completed. Replanning was aimed both at how to obtain more information to diagnose the problem as well as how to protect the affected systems. It also required finding a way to resolve the conflicting goals of maximizing the safety of the systems as well as determining as confidently as possible the source of the anomaly. The team decided to alter the order in which the auxiliary power units (APUs) were shut down to obtain more diagnostic information. This change in the mission plan was then communicated to the astronauts.

After the initial assessment, responses, and communications, the new assessments of the situation and the changed plans were communicated to other controllers who were or might be affected by these changes. This happened in parallel because other personnel could listen in on the voice loops and overhear the previous updates provided to the flight director and the astronauts about the hydraulic leak. After the changes in immediate plans were communicated to the astronauts, the controllers responsible for other subsystems affected by the leak and the engineers who designed the auxiliary power units contacted the mechanical systems controllers to gain further information. In this process, new issues arose, some were settled, and sometimes issues that were previously handled needed to be re-visited or they re-emerged as new information arose.

For example, a series of meetings between the mechanical systems controllers and the engineering group were set up. This series of meetings served to assess contingencies and to decide how to modify mission plans such as a planned docking with the MIR space station, as well as re-entry. In addition, they provided opportunities to detect and correct errors in the assessment of the situation, to calibrate the assessments and expectations of differing groups, and to anticipate possible side effects of changing plans.

Additional personnel were called in and integrated with others to help with the new workload demands and to provide specialized knowledge and expertise. In this process, the team expanded to include a large number of agents in different places acting in a variety of roles and teams, all coordinating their efforts to produce a new mission and re-entry plan (from Woods & Patterson, 2000; see Watts et al., 1996 for more on the case).

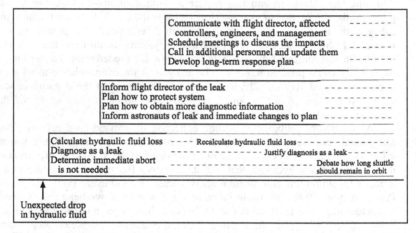

Figure 7. Schematic of an episode of anomaly response in space shuttle mission control (From Woods & Patterson, 2000).

CONTROL CENTERS IN ACTION

CSE began with studies of control centers in action (Figure 8). These studies focused on process control settings and, in particular, nuclear power plant control centers (Figure 4). The Three Mile Island accident in 1979 presented a "fundamental surprise" (Woods et al., 1994) that triggered international efforts to re-examine how teams of operators handle these kinds of plant emergencies (Rasmussen & Rouse, 1981) and to re-design analog control centers using digital technologies in the hope of providing better support for operator work (Woods et al., 1987).

These studies examined one major class of work performed by (JCSs)— *anomaly response*. In anomaly response, there is some underlying process, an engineered or physiological process which will be referred to as the *monitored process*, whose state changes over time. Faults disturb the functions that go on in the monitored process and generate the demand for practitioners to act to compensate for these disturbances in order to maintain process integrity—what is sometimes referred to as "safing" activities. In parallel, practitioners carry out diagnostic activities to determine the source of the disturbances in order to correct the underlying problem.

Anomaly response situations frequently involve time pressure, multiple interacting goals, high consequences of failure, and multiple interleaved tasks (Woods, 1988; 1994). Typical examples of fields of practice where dynamic fault management occurs include flight deck operations in commercial aviation (Abbott, 1990), control of space systems (Patterson et al., 1999; Mark, 2002), anesthetic management under surgery (Gaba et al., 1987), terrestrial process control (Roth, Woods & Pople, 1992), and response to natural disasters.

Figure 8. Spaceflight controllers in the front room at Mission Control (Courtesy of NASA).

When a researcher first encounters a control center and considers abnormal situations, they often may say the topic under study is "diagnosis" and imply that diagnosis involves information processing that links sets of symptoms to categories of faults (Woods, 1984; Rasmussen, 1986). This view mistakes one very specialized type of diagnostic setting—troubleshooting a broken device which has been removed from service—as if it instantiated a comprehensive model of anomaly response. It mistakes one strategy for diagnostic search (symptom-fault matching) as if it were a comprehensive view of the challenges of anomaly response as a general form of work in JCSs. As a result, studies and designs based on these assumptions have proved unable to penetrate the mask of an information processing sequence and have failed to recognize the deeper patterns of adaptation and dynamic control in anomaly response (*JCS-Foundations*, Chapter 1).

Work based on CSE approached the questions about how to improve operator response in abnormal and emergency situations by following the process schematicized in Figure 1:

• Making diverse observations of the interplay of people and technology in demanding work across different settings,
• Abstracting patterns across specific settings and situations to escape from surface variability—even to recognize that anomaly response is a general type of work for JCSs,
• Generating models of what makes anomaly response difficult (vulnerabilities) and what is the basis for expert and successful performance,
• Innovating promising hypotheses about what would be useful based on understanding how artifacts provide affordances relative to these demands.

What did these processes of functional synthesis reveal about anomaly response?

Cascading Effects

The driving constraint in anomaly response is that faults produce a time series of disturbances along lines of functional and physical coupling in the monitored process, or a **cascade of disturbances** (e.g., Abbott, 1990). The cascade of effects takes on different time courses in different circumstances. For example, an event may manifest itself immediately or may develop more slowly; a triggering event may occur pristine against a quiet background or in combination with other events in a changing context.

The cascade of disturbances captures the process by which situations move from canonical or textbook to non-routine to exceptional. Different domains may have different gradients in this process depending on the kinds of complicating factors that occur, the rhythms of the process, and consequences that may follow from poor performance.

Operational settings are typically data rich and noisy. Many data elements are present that could be relevant to the problem solver (Woods, 1995b). There are a large number of data channels (either through sensors, human reports, or direct perception) and the signals on these channels usually are changing and some changes are not significant. The raw values are rarely constant even when the system is stable and normal. The default case is detecting emerging signs of trouble against a dynamic background of signals rather than detecting a change from a quiescent, stable, or static background. The noisiness of the background can mask symptoms and can provide plausible alternatives to explaining changes and events away as part of a different pattern (e.g., Roth et al., 1992).

Note how basic characteristics of anomaly response can be used to guide problem-design for staged world studies. For example, a basic heuristic is to develop a scenario where the events and disturbances of interest occur against a moving or turbulent background where practitioners are already active in handling or adjusting to another event.

Interventions

The nature of the responses by practitioners affects how the incident progresses—less appropriate or timely actions (or too quick a reaction in some cases) may sharpen difficulties, push the tempo in the future, or create new challenges. The operational responses by automatic systems and by operators are part of the process of disturbance propagation and development over time and, therefore, a part of the episode itself.

Given the consequences of disturbances for safety and other pressing goals, operators cannot wait for a definitive diagnosis before acting to counter or correct the situation. "Safing" actions begin relatively early in the sequence of events. For example, while monitoring telemetry data during one space shuttle mission, flight controllers recognized high and increasing pressure in a fuel line connected to an auxiliary power unit (APU). At first, the controllers thought about several possible sources that could be producing the anomaly (a blockage in the line which in turn could be due to different sources—contamination blocking the line or a small leak could expose the hydrazine fuel to the freezing conditions of space and the resulting fuel freeze could block the line—with different responses). However, they saw that the pressure continued to increase and approach the limit for the fuel line. To keep the line from bursting, flight controllers asked the shuttle crew to cycle a relief valve to lower the pressure. This response action served as a **therapeutic intervention** (e.g., safing), which carried the goal of relieving the pressure regardless of what was producing the anomaly, so that the line would not burst under the growing pressure.

As illustrated in this example, actions to manage the consequences of disturbances are highly intertwined with diagnostic search. Critically, interventions taken for therapeutic purposes also function as **diagnostic interventions**; that is, the response of the monitored process to the intervention provides information about the nature and source of the anomaly (therapeutic interventions sometimes serve quite simply to "buy time" for diagnostic search). To continue the space shuttle example, practitioners followed up the action of cycling a relief valve by monitoring for the effect of this therapeutic intervention. The intervention was expected to produce a decrease in pressure. Instead, the flight controllers were surprised to observe a quite different response to opening of the relief valves. The telemetry value of the fuel line pressure continued to gradually rise. Since opening the relief valves had no effect on the pressure value, the practitioners began to hypothesize that the telemetry signature must have been due to a failed transducer or pressure sensor. In this case, the intervention that was intended to be therapeutic also became a diagnostic intervention because it provided practitioners with more evidence about possible explanations (note that actions taken only for diagnostic value, as interventions, may still have risks relative to the monitored process; e.g., checking a difficult to access area of a process may introduce new problems, or in cardiovascular anesthesiology, different actions to gain information pose varying risks of other injuries to the patient).

Revision

The time course of disturbances and interventions produces a series of changes in the data available about the state of the monitored process. Evidence about the anomaly and what produced it comes in over time. The high fuel line pressure case illustrates the process: eventually the flight controllers observed that the fuel line pressure leveled off and stayed constant (later, pressure began to drift slowly back toward normal). As they pondered what would account for this additional and unexpected behavior, the practitioners also began to consider how to modify plans for future mission events and evaluate contingencies that might arise in light of the anomaly. In working through this thread in anomaly response, the flight controllers spoke with others at the shuttle launch site and learned that the shuttle had been exposed to a rainstorm while it was waiting to launch. Therefore, the insulation surrounding the anomalous fuel line might have gotten wet. The new information, combined with the new events in pressure behavior (stabilize, then drift lower) led some practitioners to hypothesize that the high pressure was caused by wet insulation. They proposed that when the shuttle reached outer space, the insulation around the line froze and, therefore, caused the hydrazine fuel to freeze. As the ice sublimated from the insulation and the heaters warmed the line, the frozen hydrazine began to melt and return to its normal state. This new hypothesis accounted for all of the findings to be explained and therefore many practitioners involved believed that this provided the best explanation for the anomaly.

This example highlights the critical demand factor and basic vulnerability in anomaly response—**revision or failures to revise** of assessments. The evidence across studies of anomaly response show that initial explanatory hypotheses tend to be correct or plausible given the evidence available at that point. Problems arise later as the situation changes and as new evidence comes in, if the practitioners fail to revise their assessments in light of these changes. In other words, the initial assessments are basically correct (given what information is available) but practitioners can get stuck in their previous assessment and fail to revise these assessments as the situation changes. To see specific results, check the synthesis from multiple studies of operator performance in nuclear power emergencies, both observation during high fidelity simulations and retrospective analyses of actual accidents in Woods et al., 1987 and also Roth et al., 1992; or see Paul Johnson's line of research across Johnson et al., 1988; 1991; 1992; 2001).

Garden path problems are a specific class of problems where revision is inherently difficult since "early cues strongly suggest [plausible but] incorrect answers, and later, usually weaker cues suggest answers that are correct" (Johnson, Moen & Thompson, 1988). The following is an example of a garden path situation from nuclear power plant failures (from Roth, Woods & Pople, 1992).

An ISLOCA is an interfacing system loss of cooling accident in one kind of nuclear power plant. It can happen if a double valve failure occurs in the pipes that connect the high pressure cooling system to the low pressure cooling system. The most salient early symptoms of an ISLOCA can lead one down a garden path because they suggest another fault: a primary system break inside containment. The operational context encourages the garden path aspect in that once the symptoms consistent with a primary system break are observed, the goal becomes to maintain primary system integrity, placing a significant demand on operator attention and activities (potentially diverting resources from situation assessment and additional hypothesis generation and evaluation).

Since the evidence strongly indicates a primary break inside containment, during emergencies, crews are supposed to follow emergency procedures that are specific to the particular diagnostic conclusion about the type of anomaly or fault present. It is interesting to note that once the operators begin to follow the primary "break inside containment" procedure, there is no provision within the structure of the procedure system to redirect them to the ISLOCA procedure (this is a hole in how the procedure set was designed to accommodate the general risk of misdiagnosis of fault type). It is only over time that the complete set of symptoms that indicate the actual nature of the fault build up, making the primary system break to containment diagnosis increasingly less plausible.

Garden path problems are one category of problems that illustrates how anomaly response is fraught with inherent uncertainties and trade-offs. Incidents rarely spring full blown and complete; incidents evolve. Practitioners make provisional assessments and form expectancies based on partial and uncertain data. These assessments are incrementally updated and revised as more evidence comes in. Furthermore, situation assessment and plan revision are not distinct sequential stages, but rather they are closely interwoven processes with partial and provisional plan development and feedback that lead to revised situation assessments (Woods, 1994; see Figure 11, p. 86).

As a result, it may be necessary for practitioners to entertain and evaluate those assessments that later turn out to be erroneous. In the garden path problem above, it would be considered quite poor performance if operators had not seriously considered the primary break to containment hypothesis or if they had been too quick to abandon it since it was highly plausible as well as a critical anomaly if present. The vulnerability lies not in the initial assessment, but in whether or not the revision process breaks down and practitioners become stuck in one mindset or even become fixated on an erroneous assessment, thus missing, discounting, or re-interpreting discrepant evidence. Unfortunately, this pattern of **failures to revise** situation assessment as new evidence comes in has been a part of several major accidents (e.g., the Three Mile Island accident; Kemeny, 1979).

Fixation

Cases of fixation begin, as is typical in anomaly response, with a plausible or appropriate initial situation assessment, in the sense of being consistent with the partial information available at that early stage of the incident. As the incident evolves, however, people fail to revise their assessments in response to new evidence, evidence that indicates an evolution away from the expected path. In fixations, the practitioners appear to be stuck on the previous assessment. They fail to revise their situation assessment and plans in a manner appropriate to the data now present (for data on fixations see De Keyser & Woods; 1990; Johnson et al., 1981; 1988; 2001; Gaba et al., 1987; Rudolph, 2003). A failure to revise can be considered a **fixation** when practitioners fail to revise their situation assessment or course of action and **persist** in an inappropriate judgment or course of action **in the face of opportunities to revise**.

One critical requirement in order to describe an episode as a fixation is that there is some form of **persistence** over time in the behavior of the fixated person or team. This means cases of fixation include opportunities to revise, that is, cues, available or potentially available to the practitioners, that could have started the revision process if observed and interpreted properly. In part, this feature distinguishes fixations from simple cases of inexperience, lack of knowledge, or other problems that can impair detection and recovery from mis-assessments. The basic defining characteristic of fixations is that the immediate problem-solving context has framed the practitioners' mindset in some direction that is inappropriate given the actual evolution of the episode. In naturally occurring problems, the context in which the incident occurs and the way the incident evolves activates certain kinds of knowledge as relevant to the evolving incident (as described in the perceptual cycle; *JCS-Foundations*, p. 20). This knowledge, in turn, affects how new incoming information is interpreted. In psychology of cognition literature, this is often referred to as a *framing effect*. Another distinguishing mark in cases of fixation is that, after the fact, or after the correct diagnosis has been pointed out, the solution seems obvious, even to the fixated person or team.

Analyses of failures to revise that have occurred in field studies suggest several different forms of persistence (see De Keyser & Woods, 1990). In some cases the operators seem to have many hypotheses in mind, but never the correct one. The external behavior exhibited appears incoherent as they jump from one action to another one, but without any progress. Ongoing persistence on a single line of thought and action is seen in other cases, usually taking the form of multiple repetitions of the same action or the same monitoring check despite an absence of results or feedback. The operators seem to detect the absence of any effect or change, yet continue in the same vein. In a third pattern the operators seem insensitive to new cues and evidence. Cues discrepant with their model of the situation seem to be missed, discounted, or reasoned away (sensor failures and false alarms may be quite plausible reasons given past experiences).

Feltovich et al. (2001) studied pediatric cardiologists in a garden path scenario and abstracted a set of knowledge shields that people use to explain away discrepancies that could signal that their assessment or framing of the situation is mistaken. Expertise at revision appears to be quite special and seems to depend on taking a "suspicious stance" toward any data that fails to fit the current assessment, despite the relative prevalence of "red herrings" (data changes sometimes are not significant or due to other minor factors).

There is a complementary vulnerability to be avoided too, which has been called *vagabonding* (Dörner, 1983). Not every change is important; not every signal is meaningful. If the team pursues every signal and theme, they can jump incoherently from one item to the next, treating each superficially and never developing a coherent coordinated response. Shifting attention constantly from one change, item, or thread to another undermines the ability to formulate a complete and coherent picture of the trajectory of the system. Conversely, not shifting attention to new signals can result in missing cues that are critical to revising situation assessment. The contrast captures the potential trade-off between the need to revise assessments and the need to maintain coherence.

When adopting a suspicious stance, practitioners recognize the fragility and tentative nature of assessments and actively invest in continuing to check for and exercise sensitivity to new cues that could update or modify their assessment, *despite* the need (or pressures) to carry out other necessary activities consistent with their roles (Klein, Pliske, et al., 2005). For dramatic contrasting examples of the difficulty, examine the history of how mission control handled anomalies during the Apollo program as described in Murray & Cox, 1989 (especially examine Apollo 13 with regards to what mission control was doing to make sense of the many anomalous readings during the first 40 minutes following the initial presentation of the anomaly. Consider how the team avoided becoming fixated on the interpretation that the multiple anomalies were due to an instrumentation problem); then contrast this with how the Mission Management Team quickly discounted uncertain evidence of trouble rather than following up vigorously (see CAIB, 2003, Chapter 6; Woods, 2005b).

Generating Hypotheses

Performance at diagnostic reasoning depends on the ability to generate multiple plausible alternative hypotheses that might account for the findings to be explained. Observing anomaly response in action reveals that people do not call to mind all plausible alternatives at once. Rather, context, new evidence, and the response of the monitored process to interventions all serve as cues that suggest new possible explanations. Additionally, they change the relative plausibility of different candidates in ways that shift which ones guide the diagnostic search for additional evidence. In the space shuttle high fuel line pressure anomaly case, note how the set of candidate explanations shifted over time.

This part of anomaly response is referred to as hypothesis generation. Since diagnostic reasoning involves evaluation of hypotheses relative to available competitors, the generation of possible alternative hypotheses is an area that can influence the quality of reasoning. One strategy for supporting expert performance is to aid hypothesis generation, i.e., to **broaden** the set of hypotheses under consideration as candidate explanations for the pattern of findings. Research strongly suggests that (1) multiple people—in the right collaborative interplay—can generate a broader set of candidates than individuals, regardless of level of individual expertise (Gettys et al., 1979; Gettys et al., 1986; Gettys et al., 1987; Hong & Page, 2002 and (2) using automation to remind or critique during human practitioners' anomaly response process—with the right interaction design—produces a broader exploration of possibilities (and perhaps deeper evaluation as well) compared to human-machine architectures where people are asked to monitor the automation's assessments and activities (Woods, 1986a; Layton et al., 1994; Guerlain et al., 1996; Smith et al., 1997; Guerlain et al., 1999). The critical performance attribute is broader or narrower exploration in pace with the evolving situation. Supporting this is inherently a collaborative, multi-agent issue.

As we will see in exploring breakdowns in coordination (automation surprises in Chapter 10), the conventional and common belief that people should be placed in the role of monitoring the automation's assessments and activities produces surprising negative effects. Having the machine first suggest an explanation or provide an assessment can narrow the range of alternative accounts considered by the human monitor (and the total set explored by the collaborative ensemble). Studies have shown that when their evaluation is framed by the machine's suggestion, people miss weaknesses or limits in the machine's account and to accept accounts/plans that they would not find appropriate when doing the task themselves (Layton et al., 1994; Smith et al., 1997). This finding strongly illustrates the advantage of taking a JCS perspective over evaluating and designing each component in isolation. Supporting broadening and avoiding premature narrowing depends on designing the coordination between agents. These joint system issues cannot be addressed if one develops a machine advisor or automation first and then attempts to graft on a subsequent design of a human interface.

The issues of hypothesis generation and revision of assessments are all complicated by the fact that multiple faults or factors may be present and producing the observed time varying course of disturbances. In complex, highly coupled monitored processes, multi-fault or multi-factor cases may be relatively likely or disproportionately risky. For example, faulty sensor readings may be relatively likely to occur in combination with another fault when processes are heavily instrumented or prone to interference. Diagnosis difficulty increases when there is greater coupling or interactions within the monitored process, as this produces more complicated disturbance chains, which exacerbate the difficulties inherent in distinguishing when multiple factors are present (Woods, 1994). For example, disturbances arising from separate underlying conditions may interact and appear to be due to single failure. The question—Is a series of disturbances the effect of a single fault, or are they due to multiple factors?—is omnipresent in anomaly response. Other related diagnostic difficulties in anomaly response include *effects at a distance* where the symptoms that are visible or salient arise from disturbances physically or functionally distant from the actual fault.

Again, note how different aspects of multiple faults and disturbance chains can be used in problem-design for staged world studies. For example, one heuristic is to interject a smaller fault that masks a portion of the second more significant fault of interest and the disturbance chains it produces.

Recognizing Anomalies

Formal accounts of diagnosis often overlook the demands of recognizing, out of many changing data channels, which changes or lack of change are anomalies, and which are less significant (or are even distractions). The inherent variability of the physical world means that changes constantly occur in the monitored process that could be relevant, in principle.

When artificial intelligence algorithms first attempted to automate diagnosis, the programs relied on people to feed them the findings to be explained (Woods et al., 1990). Historically, system developers usually labeled these as "consultants" or as "optional advice-givers"; but they were developed in fact to move toward autonomous systems (Woods, 1986a). The initial attempts to take autonomy further required connecting these programs to sensor and telemetry data from real-time processes, but these attempts quickly broke down because, without the human mediator, the programs could not distinguish which, out of the large number of data changes, were findings-to-be-explained. In other words, the programs immediately collapsed in the face of data overload (*JCS-Foundations*, pp. 79-80). Modeling anomaly response required inclusion of a mechanism for anomaly recognition as a basic demand factor given the large number of data channels and the inherent variability of the physical and human processes being monitored (Roth et al., 1992; Woods, 1994). Discovering the critical role of anomaly recognition then provided a guide in the search for new affordances to support JCSs.

Making progress on anomaly recognition first required a distinction (Woods, 1994; Woods, Patterson, & Roth, 2002): anomalies could be about discrepancies between what is observed and what is expected, or about discrepancies between observed and desired states. The second type—abnormalities—indicated the need to act and to modify courses of action (safing, contingency evaluation, replanning). The first type are violations of expectation, and it is these which act as findings-to-be-explained and trigger lines of diagnostic reasoning.

Second, studying anomaly recognition led to the insight that control of attention was a critical part of anomaly response (Woods, 1994). Attention in multi-threaded situations is inherently about balancing a trade-off (Woods, 1995b): too many irrelevant changes can be pursued (data overload), but one can discard too many potentially relevant changes as irrelevant. In the former case, performance degrades because there are too many findings to be pursued and integrated into a coherent explanation of process state. Symptoms include loss of coherent situation assessment and vagabonding (Dörner, 1983). In the latter case, performance degrades because too many potentially relevant changes are missed, rejected as irrelevant, or discounted, increasing the danger of failures to revise assessments.

Note how in developing a functional synthesis of JCSs at work in anomaly response we continue to stumble onto surprising findings. This section adds to our list: (a) diagnosis is effected as much by deciding what is to be explained as it is about the mechanisms for generating and evaluating alternative candidates, and (b) recognizing anomalies depends on violations of expectations.

The Puzzle of Expectancies

> ... readiness to mark the unusual and to leave the usual unmarked—to
> concentrate attention and information processing on the offbeat.
> J. Bruner, 1990, p. 78

Our attention flows to unexpected events or departures from typicality (Chinn &
Brewer, 1993). This observation reveals how meaning lies in contrasts—some
departure from a reference or expected course, or differences against backgrounds
(Woods, 1995a).

An event may be expected in one context and, therefore, go apparently
unnoticed, but the same event will be focused on when it is anomalous relative to
another context. An event may be an expected part or consequence of a quite
abnormal situation, and, therefore, draw little attention. But in another context, the
absence of change may be quite unexpected and capture attention because reference
conditions are changing. Again, the simple story of the serendipitous auditory
display in the nuclear power control room (see the case on pp. 15-16) illustrates the
process. The audible clicks provided observability about the behavior of the
automated system, which allowed operators to develop and demonstrate skilled
control of attention. When the behavior of the automated system was expected in
context, the operators did not orient to the signals (or lack of clicking sounds).
Initially, outside observers could not tell that the practitioners were even noticing
the sounds. Only when the auditory display indicated activity that was unexpected
did the observers see explicit indications that the practitioners were using the
clicking sounds as an auditory display.

A series of studies of highly skilled teams of practitioners in NASA mission
control provide some insights about what practitioners find informative both before
and during anomaly response (Patterson et al., 1999; Patterson & Woods, 2001;
Chow et al., 2000). These studies found that events are highly prominent in the
different forms of communication exchange between spaceflight controllers during
missions.

For example, in an analysis of the contents of flight controllers' shift logs, Chow
(2000) found that references to change and behaviors outnumbered references to
states by about 3/1, and outnumbered references to base data values by nearly 20/1.
Similarly, Patterson & Woods (2001), in a study of shift handovers between flight
controllers noted that "practitioners rarely discussed base data values (e.g., "the
pressure is 82 psi"), but rather described data in terms of event words and phrases
that signified a temporal sequence of behaviors (e.g., "there was a water spray
boiler freeze up")." Patterson et al. (1999) analyzed the role of the "voice loops"
network over which operational groups communicate in mission control. They
noted that the communication occurring over the most heavily monitored channels
typically consisted of integrated descriptions of conditions, behaviors, and activities
in the space shuttle systems (e.g., on the flight director's loop). By simply *listening
in*, flight controllers could learn what was going on in systems outside their scope
of responsibility and, thus, *anticipate* impacts on their own systems and activities.
With respect to the behavior of the shuttle systems, the key point was that the work

of integrating patterns of base data into descriptions of operationally significant behaviors and sequences had already been done by the controllers responsible for monitoring those sub-systems. The voice loops allowed flight controllers to absorb the context of operations at an integrated, semantic, and temporal level, i.e., in terms of events.

Note the trick the study team used to see, when they didn't know exactly what to look for. The studies were built on the assumption that experienced practitioners generally are competent meaning-recognition systems (Flach et al., 2003). This led investigators to focus the conditions of observation on points and times when practitioners need to exchange information (handovers, annotations on logs, how they monitor voice loops during anomalies, communications between groups during meetings to re-plan following an anomaly). The data capture and integration then focused on what the practitioners find significant out of large and changing data sets, either in terms of what was informative to them when monitoring their peers' activities or in terms of what they found worth communicating to their peers (in handovers or as annotations on shift logs).

Given these results from natural history studies (direct observation), Christoffersen, Woods & Blike (2006) followed up with a Staged World study of what anesthesiologists found "interesting" as they monitored a standard real-time display of sensor data during a simulated surgical scenario (note how the line of research switched natural laboratories—mission control to operating room—as well as shifted from direct observation to observing in staged world simulations). Christoffersen et al. traced how multiple attending physicians and residents recognized events from the continuous flow of telemetry data and obtained verbal protocols about what the practitioners found informative at these points. The results illustrated that experts are highly sensitive to event patterns in the underlying process, such as the general deteriorate/recovery event pattern. Similar to the studies of space mission control, Christoffersen et al. found that, by a conservative measure, a dominant proportion (fully two-thirds) of the informative features identified by the participants could be classified as event-related, especially assessments of the character of events in progress and event patterns that required integration over the behavior of multiple variables.

To see the role of expectation in anomaly recognition, consider the example in Figure 9. How does one interpret the observation of a decrease in pressure (in event words—"falling")? Is this fall an event in need of explanation? The answer depends on what has preceded and what is expected to happen next. And both of those depend on the observer's model of the influences impinging on the process (i.e., the inputs driving change), and their model of the process dynamics (i.e., the constraints on how the process should behave in response to various inputs). Changing the influence model changes the interpretation of what behavior fits and what does not, and what can happen next.

The space of variations in just this simple example in Figure 9 is huge. If, prior to interval **t,** practitioners recognized that an anomaly was in progress based in part on falling pressure, and that no actions had been taken to counteract the fall, then the decrease is expected and the continued fall is expected. If the normal response to pressure drops is for an automatic system to begin to respond in order to stabilize pressure, then the same decrease can be interpreted as part of a signature of pressure leveling off or beginning to turn around—the automatic system responded as expected and is compensating for the disturbance in pressure. If the decrease has been going on and an automatic system typically responds quickly, the same decrease becomes unexpected since the typical response would have produced a change in pressure behavior. In this case, the same behavior is unexpected and would trigger further investigation: Did the automatic system fail (and can I still restart it or is it unavailable)? Is the fault different from those the automatic system can compensate for? Is the fault larger than the capacity of the automatic system? This example illustrates how what is meaningful depends on context (Woods, 1995b; Woods et al., 2002; Theureau, 2003).

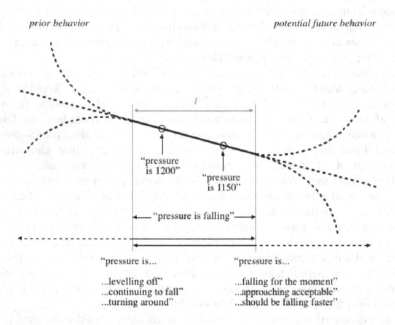

Figure 9: The same behavior in some interval **t** can be interpreted in multiple ways depending on context and expectations of the observer (From Christoffersen & Woods, 2003).

First, notice how expectations provide a way to determine what changes (or absence of changes) are informative and therefore worth practitioners' attention or worth their colleagues' attention. Using expectations to define a kind of anomalous event is essential to avoiding data overload (Woods et al., 2002). Second, expectations are defined relative to a changing model of the active influences thought to be driving the behavior of the monitored process (Figure 10).

The model of active influences may include items such as what actions have been performed on the process or what disturbances are currently at work (e.g., a malfunction in an engineered system or disease process in a medical context).

Knowledge of system dynamics serves to specify how the process ought to behave in response to the active influences. The content of these models will depend on the specific system involved, but together they combine to produce a set of more or less specific expectations and reference states against which the actual behavior of the process can be contrasted.

It is particularly important to note the critical test cases defined by the above model of anomaly recognition. One test case focuses on how practitioners would not explicitly follow-up an abnormal event with diagnostic search when the event is part of the expected sequence that follows from a fault thought to be active or part of a disturbance chain that is thought to be underway. Rather, the expected though abnormal behavior reinforces the practitioners' current assessment.

The second and more difficult test case is to recognize the absence of change as an unexpected event. Because these models of influences allow people to understand how the process should change in certain situations, it can sometimes be the case that absence of change is itself an informative event. For example, if an action is taken to alter the trajectory of a process, the model of system dynamics can be used to determine how the process should change. If the process continues to behave in the same way subsequent to the action, this signature may define a significant event.

Christoffersen et al. (2006) found evidence for this type of event in an episode involving a lack of change in the simulated patient's blood pressure, even after several doses of medication had been administered to bring pressure down. The normal onset time for the effect of the medication is relatively short, which produced an expectation that blood pressure would quickly start to decrease. The physician then could assess the size of the decrease produced by the initial amount of medication given and make further adjustments until pressure returned to normal range. However, the study participants reacted strongly when blood pressure continued to increase (no response to the intervention), even leading some to update their model of the nature or severity of the patient's underlying problem.

The above results (cf. also Teigen & Keren, 2003) have led Christoffersen & Woods (2003) to propose a functional synthesis of anomaly and event recognition from a stream of data (Figure 10). This model helps frame a number of points about why anomaly recognition is particularly challenging (cf., also Theureau, 2003). First, the synthesis shows how anomaly and event recognition is tuned to the future—what can happen next. For the high blood pressure example, multiple interventions occurred before pressure stabilized and began to turn around. Given the extra control interventions needed to overcome the disturbances, the issue shifted to the danger of overshooting the target range and heading toward low blood pressure with the need for additional interventions to bring the parameter sufficiently under control.

Second, note how, by knowing what to look for, one notices what one was not expecting to see. This is a key criterion that tests for *observability*, or feedback that provides insight into a process—the ability to notice what one was not expecting, otherwise new technology or new designs merely make data available while leaving all of the work to find and integrate the important bits to the observer. Tukey (1977, p. vi) referring to statistical graphics, expressed the target memorably:[7]

"The greatest value of a picture is when it forces us to notice what we never expected to see."

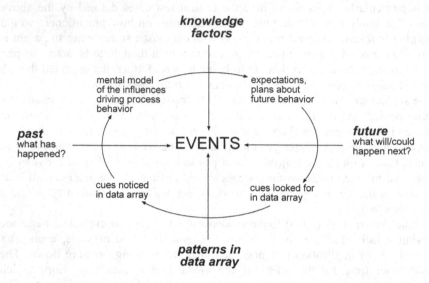

Figure 10. A model of factors underlying anomaly recognition from event patterns in a dynamic telemetry stream (From Christoffersen & Woods, 2003).

Third, expectations depend on experience by building up a highly differentiated set of what is typical behavior for different contexts (and this capability probably has more to do with processes of perceptual learning—elaborate differentiation—than with the learning mechanisms implemented in symbolic computer processing models of an individual's working memory).

In addition, what is informative goes beyond the literal surface properties of the data. As Woods has emphasized (1995a; Woods et al., 2002), what is informative depends on the behavior of the data array relative to the mindset of some observer—goals, knowledge, expectations (cf. the concept of mutuality from Gibson, 1979, discussed on p. 7). Christoffersen et al. (2006) demonstrated this in their event recognition study as the events marked by experts often did not track with simple changes in the display of a single parameter (though for observers with less expertise, the events noted more closely tracked to simple surface changes).

---

[7] The same concept is also central in Rasmussen's (1986) notion of topographic search in diagnosis.

Since the definition of what is informative depends on properties of the observer and not just properties of the world, in any given interval, people with different mindsets may isolate entirely different sets of events and focus on different anomalies from the same stimulus stream. As Ginsburg & Smith (1993) have observed in the context of social perception, the potential for divergent (but still valid) interpretations of the same stimulus stream tends to increase as the level of analysis shifts from low-level physical transitions (is pressure increasing) to event patterns (absence of a response to an intervention) to larger event sequences (a disruption to a plan in progress). Different observers may agree on the set of low-level physical changes that have occurred, but vary drastically in terms of the higher-level events that they perceive, depending on their particular mindset. Hence, the process tracing described in Chapter 5 centers on tracing how mindset is created, then shifts, or fails to shift, as situations evolve.

The temporal scale of events varies as events come nested at multiple levels of analysis (Warren & Shaw, 1985). That is, changes in stimuli defined over any given temporal interval may provide information about different events that are occurring simultaneously at widely varying timescales. For example, data indicating deteriorating performance in a subsystem onboard the space shuttle may have immediate operational significance (e.g., indicating a need to switch to a backup system to compensate for the short term). But the same event may be part of a larger pattern on the scale of the mission as a whole (e.g., at this scale the anomaly might be a loss of capability which creates an impasse that blocks execution of the current mission plan and requires replanning), or for the entire program (e.g., at this scale the anomaly might represent a long term re-design issue across the fleet that challenges the overall shuttle launch and payload schedule for months).

Typically, there is no single privileged timescale; the relevant level of analysis is a function of the observer and the context. In other words, events and anomalies are inherently systems issues (and thus, there is grave doubt whether events can be processed solely through mere data manipulation by automata) in all three of the senses that are critical to distinguishing any systems engineering (and therefore in systems engineering on cognitive systems): emergent relationships, cross-scale interactions, and sensitivity to perspective (see p. 7 in Chapter 1).

Control of Attention

> Everyone knows what attention is. It is the taking possession by the mind, in a clear and vivid form, of one out of what seem several simultaneously possible objects or trains of thought.
> William James, 1890, I, p. 403-404

The functional synthesis of anomaly response so far has revealed how practitioners make provisional assessments and form expectancies based on partial and uncertain data. These assessments are incrementally updated and revised as more evidence comes in, as the disturbances cascade, and as interventions occur to contain the effects of these disturbances. Situation assessment and plan revision/action are not distinct sequential stages, but rather they are closely interwoven processes with partial and provisional plan development and feedback leading to revised situation assessments. This means anomaly response inherently demands a multi-threaded process as indicated in Figure 11 (which is another elaboration on Neisser's perceptual cycle; *JCS-Foundations*, p. 20).

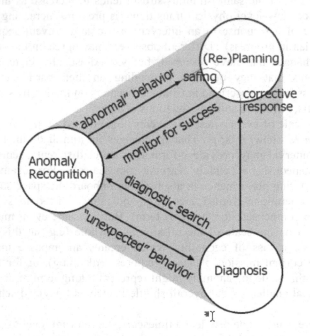

Figure 11. The multiple intermingled lines of reasoning in anomaly response (from Woods, 1994).

Practitioners need to continue to monitor for new changes; they need to look for specific information to support or disconfirm possible explanations of the unexpected findings; they need to follow up checking whether interventions produce the desired and expected results. Abnormalities require interventions, contingency evaluation and replanning activities. Anomaly response involves skillful juggling of these multiple threads. As a multi-threaded activity, anomaly response places significant demands on control of attention.

Focus of attention is not fixed, but shifts to explore the world and to track relevant changes in the world. On flight decks, in operating rooms, and in everyday work activities, attention must flow from object to object and topic to topic. In other words, one re-orients attentional focus to a newly relevant object or event from a previous state where attention was focused on other objects or on other cognitive activities (such as diagnostic search, response planning, and communication to other agents). New stimuli are occurring constantly. Sometimes such new stimuli are distractions. But other times, any of these could serve as a signal that we should interrupt ongoing lines of thought and re-orient attention.

This re-orientation involves disengagement from a previous focus and movement of attention to a new focus (Gopher, 1992).

We are able to focus, temporarily, on some objects, events, or actions in the world or on some of our goals, expectations or trains of thought *while remaining sensitive to new objects or new events that may occur*. It is the last clause that is the critical feature for control of attention—the ability to refocus without unnecessary disruptions to ongoing lines of work.

Thus, a basic challenge for any agent at work is *where to focus attention next in a changing world*. Which object, event, goal or line of thought we focus on depends on the interaction of two sets of activity as specified in Neisser's perceptual cycle (*JCS-Foundations*, p. 20). One of these is goal or knowledge directed, endogenous processes that depend on the observer's current knowledge, goals and expectations about the task at hand. The other set of processes are stimulus or data driven where attributes of the stimulus world (unique features, transients, new objects) elicit attentional capture or shifts of the observer's focus. These salient changes in the world help guide shifts in focus of attention or mindset to relevant new events, objects, or tasks.

The ability to notice potentially interesting events and know where to look next (where to focus attention next) in natural perceptual fields depends on the *coordination* between orienting perceptual systems (i.e., the auditory system and peripheral vision) and focal perception and attention (e.g., foveal vision). The coordination between these mechanisms allows us to achieve a "balance between the rigidity necessary to ensure that potentially important environmental events do not go unprocessed and the flexibility to adapt to changing behavioral goals and circumstances" (Folk et al., 1992, p. 1043).

The orienting perceptual systems function to pick up changes or conditions that are potentially interesting and play a critical role in supporting how we know where to look next. To intuitively grasp the power of orienting perceptual functions, try this thought experiment suggested by Woods & Watts (1997): put on goggles that block peripheral vision, allowing a view of only a few degrees of visual angle. Now think of what it would be like to function and move about in your physical environment with this handicap. Perceptual scientists have tried this experimentally through a movable aperture that limits the observer's view of a scene (e.g., Hochberg, 1986). Although these experiments were done for other purposes, the difficulty in performing various visual tasks under these conditions is indicative of the power of the perceptual orienting mechanisms.

Alarms and Directed Attention

Alarm Overload
"1202." Astronaut announcing that an alarm buzzer and light had gone off and the
        code 1202 was indicated on the computer display.
"What's a 1202?"
"1202, what's that?"
"12...1202 alarm."

Mission control dialog over voice loops as the LEM descended to the
moon during Apollo 11 (Murray & Cox, 1989).

"The whole place just lit up. I mean, all the lights came on. So instead of being able to
tell you what went wrong, the lights were absolutely no help at all."

Comment by one space controller in mission control after the Apollo
12 spacecraft was struck by lightning (Murray & Cox, 1989).

"I would have liked to have thrown away the alarm panel. It wasn't giving us any
useful information."

Comment by one operator at the Three Mile Island nuclear power
plant to the official inquiry following the TMI accident (Kemeny,
1979).

"When the alarm kept going off then we kept shutting it [the device] off [and on] and
when the alarm would go off [again], we'd shut it off." "... so I just reset it [a device
control] to a higher temperature. So I kinda fooled it [the alarm] ... "

Physicians explaining how they respond to a nuisance alarm on a
computerized operating room device (Cook, Potter, Woods &
McDonald, 1991).

The present lunar module weight and descent trajectory is such that this light will
always come on prior to touchdown [on the moon]. This signal, it turns out, is
connected to the master alarm—how about that! In other words, just at the most
critical time in the most critical operation of a perfectly nominal lunar landing, the
master alarm with all its lights, bells and whistles will go off. This sounds right lousy
to me. ... If this is not fixed, I predict the first words uttered by the first astronaut to
land on the moon will be "Gee whiz, that master alarm certainly startled me."

Internal memo trying to modify a fuel warning alarm computation and
how it was linked to the alarm handling system for the lunar landing
craft (Murray & Cox, 1989).

A [computer] program alarm could be triggered by trivial problems that could be
ignored altogether. Or it could be triggered by problems that called for an immediate
abort [of the lunar landing]. How to decide which was which? It wasn't enough to
memorize what the program alarm numbers stood for, because even within a single
number the alarm might signify many different things. "We wrote ourselves little
rules like 'If this alarm happens and it only happens once, don't worry about it. If it
happens repeatedly, but other indicators are okay, don't worry about it.' " And of
course, if some alarms happen even once, or if other alarms happen repeatedly and the
other indicators are not okay, then they should get the LEM [lunar module] the hell
out of there.

Response to discovery of a set of computer alarms linked to the
astronauts displays shortly before the Apollo 11 mission (Murray &
Cox, 1989).

"I know exactly what it [an alarm] is—it's because the patient has been, hasn't taken
enough breaths or—I'm not sure exactly why."

Physician explaining one alarm on a computerized operating room
device that commonly occurred at a particular stage of surgery (Cook
et al., 1991).

The above are short samples of problems with alarm systems (Norros & Nuutinen, 2005), each of which points to families of story lines about affordance and clumsiness given the demands on JCSs. This set illustrates many of the patterns on how alarms can be designed that fail to support skill at control of attention in a JCS (Woods, 1995b). Most straightforwardly, alarms are mis-designed in terms of perceptual functions: Are the perceptual signals that mark abnormal conditions in a monitored process discriminable from the background and each other (e.g., Patterson 1990)? Another perceptual function of alarms is the capacity of the signal itself to attract people's attention (exogenous control of attention or alerting). This alerting potential refers to characteristics of the signal (e.g., sudden motion or sudden sound bursts against a quiescent background) to force, to a greater or lesser degree, an observer to orient to the signal itself. Apprehending these perceptual signals conveys the message that some event has occurred which may be of interest.

As support for control of attention, alarms often are mis-designed in terms of *informativeness* of the signal that something is wrong in the monitored process: Do the semantics of alarm messages inform observers about the kinds of abnormalities present or help discriminate between normal and abnormal states or between different kinds of abnormal conditions? There are several ways that alarms can be uninformative. Alarm signals and messages may be underspecified and ambiguous as in the case of the "1202" message and in several other of the above examples.

Another reason alarms may be uninformative is the context sensitivity problem. Often, what is called an "alarm" actually indicates the status of a parameter, subsystem, or component. However, current status may or may not be actually abnormal without reference to additional data which specifies the relevant context. The discussion of what the various computer alarms mean in the case above from Apollo 11 serves as a vivid account of the general problem of context sensitivity for finding what is meaningful and escaping from data overload (also see Figure 9; Woods et al., 2002).

Sometimes alarms occur frequently and seem so familiar that observers miss the cases where something truly unusual is happening (Xiao et al., 2004). In other words, the number of false alarms is high relative to the number of times the alarm signals that the abnormal condition it monitors actually is present (Sorkin & Woods 1985). The counter-intuitive result, established through these mathematical models (and follow-up empirical studies) of the impact of false alarm rates, is that increases in false alarm rates rapidly reduce the inherent informativeness of alerting signals. For example, the statistic "positive predictive value" (PPV = [True positive] / [True + False positives]) is one measure that can begin to establish whether an alarm signal has any information to convey (Getty et al., 1994).

Alarms often are mis-designed in a third way: as a collaborative agent that attempts to re-direct the attention of another—the human observer (Woods, 1995b). Directed attention is a kind of coordination across agents where one agent can perceive and direct the attentional focus of other agents to particular parts, conditions, or events in the monitored process. Directing another's attention and perceiving where another agent's attention is directed are basic aspects of human function as social and cognitive agents (an example of how cognition is public and distributed; see Hutchins 1995a). Alarms can be seen as messages from one agent, a first stage monitor, to another, a second stage supervisory agent who monitors multiple channels and whose cognitive and physical resources can be under moderate to severe workload constraints. The alarm message is an interrupt signal intended to re-direct the attention of the supervisor from their ongoing activity to examine some particular area or topic or condition in the monitored process. In effect, the attention directing signal says, "There is something that I think that you will find interesting or important; I think you should look at this." Thus, alarm systems participate in processes of *joint reference* as signals refer to events in the monitored process in order to direct another's attention to an event (Bruner 1986).

The receiver must use some partial information about the attention directing signal and the condition that it refers to, in order to "decide" whether or not to interrupt ongoing activities and lines of reasoning. Some attention directing signals should be ignored or deferred; similarly, some attention directing signals should re-direct attention. The quality of the control of attention is related to the skill with which one evaluates interrupt signals without disrupting ongoing lines of reasoning—knowing when the attention directing event signals "important" information and when the attention directing event can be safely ignored or deferred, given the current context.

Overall, the situation can be expressed in a signal detection theory framework (e.g., Sorkin & Woods 1985) where one can err by excessive false shifts or excessive missed shifts. Thus, in principle, two parameters are needed to describe attentional control: a *sensitivity* parameter that captures variations in skill at control of attention and a *criterion* parameter that captures tradeoffs about the relative costs and benefits of under-shifts versus over-shifts of attention. Note that framing the problem in terms of signal detection points out that even very sensitive systems for control of attention will show errors of false shifts or missed shifts of attention. Even more difficult, though, is the paradox at the heart of directed attention. Given that the receiving agent is loaded by various other demands, how does one interpret information about the potential need to switch attentional focus without interrupting or interfering with current tasks or lines of reasoning already under attentional control: How can one skillfully ignore a signal that should not shift attention within the current context, without first processing it—in which case it hasn't been ignored?

Alarms systems are rarely explicitly designed to function effectively as attention re-directors or as part of a distributed system of agents that coordinate activities as situations evolve. But go back to the case of the serendipitous auditory display in the nuclear power control room (pp. 15-16) for a successful example of the required affordance. Note the key contrast in this case that illustrates the paradox in directed attention—when the auditory display provided information that was unexpected the operators shifted attention to that part of the monitored process, but when the auditory signals were expected given ongoing conditions there were no explicit signs that the operators heard anything at all.

A study by Dugdale et al. (2000) of human–human coordination in an emergency call center illustrates the processes involved in directed attention across agents under workload. They found that directing another's attention depended on being able to see what the other agent is doing in order for one agent to be able to judge when another was interruptible. In other words, interruptibility is a joint function of the new message and the ongoing activity. This requires that one agent is able to see the activity of the other in enough detail to characterize the state of the other's activities—what line of reasoning are they on? Are they having trouble? Does their activity conform to your expectations about what the other should be doing at this stage of the task? Are they interruptible? This kind of coordination illustrates how joint reference works in two directions. In one direction an agent signals another by "referring to something with the intent of directing another's attention to it" in a mentally economical way (Bruner 1986, p. 63). In the other direction, one agent can perceive where and to what another is directing their attention, without requiring any explicit communication on the part of either agent (and the associated workload burdens). Hutchins' analysis of the joint cognitive system involved in speed control during a descent in air transport aircraft illustrates this aspect of joint reference (Hutchins 1995b). The "openness" of the representations of aircraft behavior in terms of speed management allow the physical activities associated with tasks carried out by one agent to be available for the other pilot to pick up without requiring explicit intent to communicate and without disrupting either's ongoing lines of reasoning (see also, Norman 1990).

To coordinate in processes of joint reference depends on the external representations of the monitored process by which agents assess the state of and interact with the monitored process. One refers to some part, condition, event, or task in the referent world through some shared external representation of the monitored process. As a result, one agent can direct or pick up another's focus/activity on some part, condition, event or task in the referent world in mentally economical ways.

While examining the role of alarms in anomaly response reveals stories of affordances for control of attention in multi-threaded situations, in the telling of these stories we find that the storyline naturally shifts revealing the various aspects of coordination across agents that goes on in JCSs at work.

Updating Common Ground When a Team Member Returns during Anomaly Response

The following episode of anomaly response illustrates many of the processes discussed above: cascading disturbances, interventions, expectations, events, revision, reframing, control of attention, multiple threads. In addition, these processes explicitly occur in the context of multiple agents (in this case, the interactions of two individuals in the anesthesia team; the surgical team and the anesthesiology team connections). The case is taken from an observational study of expertise in anesthesia in the operating room (Johannesen, Cook & Woods, 1994). The example also is valuable as a sample of data to be analyzed in process tracing studies of JCSs at work. As a result, we provide a complete transcript of the interactions during the episode in Appendix A.

The Case

An Attending Anesthesiologist and a senior Resident Anesthesiologist together begin a neurosurgery case to clip a cerebral aneurysm. The episode of interest begins after this case is in a maintenance phase (this type of surgery is a long process with long quiet periods for the anesthesiologists). It is about an hour after induction and before the surgeons have exposed the aneurysm. The senior resident is the only anesthesiologist present; the attending has been away for about half an hour to check on another case.

An anomaly occurs and is quickly recognized against the quiet background. The senior resident sees heart rate fall (bradycardia; the opposite of this state is rapid heart rate or tachycardia), and takes corrective action by administering atropine, a drug that raises heart rate. He has the attending paged. He mentions the event to the surgeons and enquires whether they might have been doing anything related. They answer no.

To a practitioner, the bradycardia event is quite observable. The pulse rate as indicated by the beeping of the pulse oximeter suddenly slows down. The resident, who has bent down (apparently to check the urine output or to begin a cardiac output measurement), immediately gets up to look at the vital signs on the computer monitor.

Because of its severity it is critical to treat the bradycardia immediately, before its consequences begin to propagate. Five seconds later he injects the atropine. The basic action to manage the disturbance is quick and direct. It is also important to find the source of the disturbance because the base problem may lead to other disturbances (fighting each disturbance one at a time is not likely to be the best way to manage the patient's cardiovascular system).

The bradycardia is quite unexpected given what would be typical in this situation and context, though there are a variety of situations where it would not be surprising. For example, some anesthetic medications that are used for anesthetic management during surgery (i.e., during the maintenance phase) can result in a lower than normal heart rate, but the resident knows none of these influences is ongoing. Indications of low heart rate can be due to problems with the monitoring equipment, but the anomalous event in this case is too severe.

Upon the attending's return, this exchange occurs:

Attending Anesthesiologist: {re-enters operating room mid-case}
"Nice and tachycardic."

Resident Anesthesiologist:
"Yeah, well, better than nice and bradycardic."

Updating a Shared Frame of Reference

The situation calls for an update to the shared model of the case and its likely trajectory in the future. The background is a shared expectation that this is a quiet period of the case (about an hour after induction and before the surgeons have exposed the aneurysm). The workspace and information displays are "open," that is, open for view to any operationally knowledgeable party standing in the physical space. The attending, on entering the operating room, looks at the displays and notices the anomalous parameter—heart rate is high (tachycardia) and states the observed anomaly aloud (the statement is an invitation for an explanation of how the monitored process arrived in this situation). The response summarizes the physiological storyline—the patient became bradycardic, the resident noticed and responded, though the intervention produced an over-swing in the other direction (the comments also contain the shared model of the situation—slow heart rate is worse for this patient in this surgery). The exchange is very compact and highly coded, yet it serves to update the common ground previously established at the start of the case.

The resident goes on retelling the story of the event (including an event prior to the anomaly), context, and follow-up activities (actions taken and how rate responded). The update to the attending physician is full of event words: "down to and then back up", "dropped down to ...", "so I kicked her up to ..." (cf., the similar data in Patterson & Woods, 2001). The story provides information about the dynamics of the antecedent event, of the temporal flow of the event itself, and the behavior of another potentially relevant parameter (severe hypertension could produce bradycardia by a reflex pathway, but the update points out the absence of high blood pressure which rules out this mechanism for both). He mentions what action he was taking and what actions others (the surgical team) were doing ("nothing") while the event occurred, again, relating interventions (or their absence) and responses as part of the event description.

Note that this update seems to meet one general criterion for successful coordination (Patterson & Woods, 2001)—that the incoming agent can act subsequently as if they had been present throughout the evolution of the situation.

The update is also the opening to hypothesis exploration as the two anesthesiologists jointly seek explanations for the unexpected disturbance in heart rate, while at the same time the team manages cardiovascular functions to keep them within target ranges. Interestingly, the resident and attending after the update appear to be without candidate explanations as several possibilities have been dismissed given other findings (the resident is quite explicit in this case. After describing the unexpected event, he also adds—"but no explanation"). The attending tries to generate a range of hypotheses, but none seems promising— Surgeon activities? Other medications being given? Patient physiology (contractility)? Consequences of injuries from the interventions needed to place sensors for monitoring cardiovascular function ("Lines went in perfectly normal")? Low potassium?

The attending considers his experiences and points out sources that usually produce this kind of event signature (it is also important to point out that heart rate has stayed stable now with no subsequent changes)—reflexes triggered by activities associated with the surgical steps. The resident then re-examines the context the anomaly occurred in, "revisiting" what, based on the Attending's knowledge, seems to be an important time frame. As he goes back in detail over what was occurring then—they both recognize that the surgeons were engaged in an activity that could have given rise to the event—the anomaly arose at a point where the surgeons can place traction on the outermost of the three membranes (meninges) covering the brain (the Dura mater). This resolves the unexpected anomaly by providing an explanation consistent with the findings, and this explanation updates their model of influences ongoing (the source being the reflex mechanism triggered by normal surgical activities, which is an isolated factor).

While the anesthesia attending-resident coordination illustrates building common ground during an update, the anesthesia-surgical interaction illustrates a breakdown in common ground (Clark & Brennan, 1991). When the resident recognizes the anomaly, the surgeons' activities do not appear related, and when explicitly asked ("… been doing anything"), the surgeons had answered that they had not ("No"). The surgical team's response was to say that, from their perspective, nothing "unusual" was going on, which was true. The surgical team did not understand the perspective behind the anesthesia resident's question—responding to an unexpected anomaly, and his line of reasoning—searching for candidate explanation. Hence, the cross-check failed (Patterson et al., 2004).

This case illustrates the interplay of multiple parties, each with partial information, overlapping knowledge and different kinds of expertise relative to the situation at hand. The case also illustrates how different perspectives play roles in revising situation assessments. The case captures the processes of *updating* as the resident calls in additional expertise (including knowing when and how to bring additional expertise in) and is able to provide concise reconstruction of events and responses that led up to the current state during the update (Patterson & Woods, 2001).

PATTERNS IN ANOMALY RESPONSE

At this point we have introduced most of the dynamic processes that intermingle in anomaly response (Figure 11 provides a composite diagram). By grounding research on observations of the phenomena of interest, anomaly response is revealed to be quite different in almost all ways from classic assumptions that diagnosis can be isolated in terms of mapping symptoms to diagnostic categories.

For example, readers familiar with abduction as a form of reasoning will recognize that the above description of anomaly response seems consistent with models of abduction (Peirce, 1955; Josephson & Josephson, 1994). Formalist approaches to abduction debate different set covering algorithms—how to evaluate the mapping between different findings and different hypotheses—and focus on criteria for judging good coverage or "best" explanation (e.g., how to operationalize criteria that define what is parsimony in an explanation). While the results on anomaly response indicate a relationship to abductive reasoning, note the surprises that result when one starts with patterns abstracted from observation. Our tour of anomaly response has revealed a variety of critical constraints on this form of work:

- Properties of how situations evolve and present themselves,
- The knowledge that interventions occur in parallel with diagnostic search, both simplifying and complicating situation assessment,
- Avoiding data overload by being able to recognize out of a large, changing data set the findings to be explained (including the role of expectations and event patterns in recognizing anomalies),
- The demands for control of attention in a multi-threaded process,
- The knowledge that revising assessments is the key to expert performance,
- Broadening the set of hypotheses considered, even as information and assessments change,
- How replanning is connected to diagnostic assessment.

To summarize: Anomaly Response

Anomaly response is the general form of work for JCSs, and the chapter provides a functional synthesis of the basic demands in this class of work derived from converging methods observations across multiple fields of practice. Synthesizing and abstracting patterns in anomaly response provides a set of recurring story lines about resilience, coordination, and affordance. The discovery of these patterns provides a case study on understanding JCSs at work.

Additional resources: Murray & Cox's (1989) history of mission control during the Apollo project is a compact resource that captures how control centers are adapted to the demands of anomaly response. Additional studies of how mission control works are available, such as Patterson et al. (1999), Patterson & Woods (2001), Garrett & Caldwell (2002), Mark (2002), and Shalin (2005).

Chapter 9

Patterns in Multi-Threaded Work

The previous chapter considered anomaly response as one of the major forms of work in JCSs (another is modifying plans in progress). The patterns in anomaly response reveal an intermingled set of dynamic processes (Figure 11). As a general form of work, the cognitive task synthesis of anomaly response points out more general classes of demand factors that influence JCSs at work. When studying a JCS, investigators can bring these patterns to bear to facilitate the converging studies in functional synthesis. These general demand factors include tempo, escalation, coupling, premature narrowing, reframing, dilemmas, and over-simplifications.

MANAGING MULTIPLE THREADS IN TIME

Mindset is about attention and its control as evolving situations require shifts in where attention is focused in order to manage work over time, and as there are multiple signals and tasks competing for attention. Sometimes new signals are distractions from the current focus, but other times these signals are critical cues for shifts in focus. This case from Cook et al. (1992) illustrates the complexities of how mindset shifts (see also Woods et al., 1994; Dekker, 2002).

Unexpected hypotension in anesthetic management during surgery

During a coronary artery bypass graft procedure, an infusion controller device used to control the flow of a sodium nitroprusside (SNP) to the patient delivered a large volume of drug at a time when no drug should have been flowing. Five of these microprocessor-based devices, each controlling the flow of a different drug, were set up in the usual fashion at the beginning of the day, prior to the beginning of the case. The initial part of the case was unremarkable. Elevated systolic blood pressure (>160 torr) at the time of sternotomy prompted the practitioner to begin an infusion of SNP. After starting the infusion at 10 drops per minute, the device began to sound an alarm. The tubing connecting the device to the patient was checked and a stopcock (valve) was found closed. The operator opened the stopcock and restarted the device. Shortly after restart, the device alarmed again. The blood pressure was falling by this time, and the operator turned the device off. Over a short period, hypertension gave way to hypotension (systolic pressure <60 torr). The hypotension was unresponsive to fluid challenge but did respond to

97

repeated injections of neosynephrine and epinephrine. The patient was placed on bypass rapidly. Later, the container of nitroprusside was found to be empty; a full bag of 50 mg in 250 ml was set up before the case.

The physicians involved in the incident were comparatively experienced device users. Reconstructing the events after the incident led to the conclusion that the device was assembled in a way that would allow free flow of drug. Initially, however, the stopcock blocked drug delivery. The device was started, but the machine did not detect any flow of drug (because the stopcock was closed) and this triggered visual and auditory alarms (low or no flow). When the stopcock was opened, free flow of fluid containing drug began. The controller was restarted, but the machine again detected no drip rate, this time because the flow was a continuous stream and no individual drops were being formed. The controller alarmed again with the same message that had appeared to indicate the earlier no flow condition. Between opening the stopcock and the generation of the error message, sufficient drug was delivered to substantially reduce the blood pressure. The operator saw the reduced blood pressure, concluded that the SNP drip was not required, and pushed the control button marked "off." This powered down the device, but the flow of drug continued. The blood pressure fell even further, prompting a diagnostic search for sources of low blood pressure. The SNP controller was seen to be off. Treatment of the low blood pressure itself commenced and was successful.

The case is interesting because performance was both successful—managing the disturbance to protect the patient's physiology—and unsuccessful—unable to determine that the infusion device was the source of the hypotension (see also Cook et al., 1998).

As many interwoven events were happening, practitioner mindset shifted as attention flowed to some parts of the situation but not to others. Even though the practitioners did not understand the source of the anomaly, they acted quickly to correct the physiologic, systemic threat (safing).

Why didn't the infusion device receive attention to determine if it was the source of the unintended flow of drug? Severe limits on the observability of the device's activity were critical (Moll van Charante et al., 1993): the device display shows only demanded rate, not actual; misleading alarm messages; visual inspection of the device state was blocked (an aluminum shield surrounded the fluid bag, hiding its decreasing volume; the drip chamber was obscured by the machine's sensor; the clamping mechanism was hidden inside its assembly; the complexity of tubing pathways with multiple infusion devices). The belief that the device had been designed to handle all contingencies and failure modes was illusory since the design did not effectively address the potential for misassembly; the device had the potential to fail active as in this case; the sensing mechanism could not detect free flow. This overconfidence, combined with very low observability of the interface, meant that there was little in the design to support resilience of the joint system.

The practitioners reported that they turned the device off as soon as the blood pressure fell, at about the same moment that the device alarmed a second time. In their mindset, the device was off, unimportant and outside their focus on the anomaly. This does not mean that the device was examined and found to be uninvolved in the evolving situation. If this had been the case, attention would have flowed to the device for a time to investigate what it was doing, taking attention away from the other aspects of the situation.

Once the device was "off," it was dismissed from their mindset as attention flowed to bringing blood pressure under control and considering what physiological processes could be leading to the hypotension. The practitioners did not make inferences about how the device was working or not working or how it might be playing a role: they did not attend to the device at all. Once it was turned "off" the device disappeared from practitioner attention, not becoming the focus of attention again until the very end of the sequence. The device was absent from the mindset of the practitioners. This is not to say that the practitioners were idle or inattentive, indeed they engaged in looking elsewhere for sources of low blood pressure and carrying out the activities to manage that anomaly.

But "off" and a blank screen did not mean the device could not deliver drug to the patient—the "on-off" push button powers down the device; powering down the device does not guarantee the tubing is clamped shut, fluid flow is possible.

In hindsight it is easy to say if only this aspect of the situation or that aspect of device design were different the near miss would have been avoided. If the device had been more robustly designed, users might notice when situations could be near the boundaries of the assumptions about the device capabilities and the situations that can arise. If the practitioners had more experience and better feedback about how the device could act differently than intended, they might have examined the device as a (potentially) active influence in the situation. If the situation had been less fast paced, practitioners may have been able to dig out that the device was active though "off". But the point of the case is to shift our focus as researchers and designers to the challenge of supporting skilled control of attention.

TEMPO

Understanding how disturbances cascade represents a part of analyzing how external events pace practitioner activities. Note how periods of event-driven or externally paced activities can be intermingled with periods of more self-paced activity where the practitioners have control of how to invest their resources and efforts over tasks (Hollnagel, 2002). For example, space shuttle operations have high-tempo periods tied to launch, docking, or entry, but in between, depending on the kind of anomaly and the mission plan, orbit is a phase where there is a degree of more self-paced activity for flight controllers and mission engineers.

Discovering how a JCS works depends on understanding the different rhythms that play out at different scales of operation and comprehending how periods of higher tempo, more event-driven operations are intermingled with periods of lower tempo, more self-paced activity. To do this, observation and functional modeling are sensitive to what are leading and lagging factors in these dynamic processes (e.g., triggering events that produce a subsequent cascade of effects).

When we see tempo as a fundamental aspect of JCS at work, we see the fundamental demand on how to keep pace with (or stay ahead of) the changing situation. When we use this question as a way to abstract observations from multiple natural laboratories, the role of anticipation in expert performance comes to the fore. The mystery for researchers is to understand and support how expertise is tuned to the future, while, paradoxically, the data available is about the past.

The target can be captured in a principle that could be called *Avery's Wish* (Woods, 2002). Avery is an actual high-status and high-skill practitioner who is often asked by vendor engineers and usability specialists what features he would like to see in next generation devices (the questions are framed in terms of features and capabilities of the latest technologies). If the technologists could get better information from users about what features are valuable, they would know how to prioritize feature development and introduction for future products. Finally, tired of being the quarry in the feature game, Avery one day responded, "I yearn for some thing that shows me an image of what the system and the room are going to look like, ten minutes from now."

While this response was useless to the engineers in their search for feature-based requirements, it does in fact capture a fundamental aspect of practice and points to a general direction for exploring what would be useful—innovate new forms of feedback that support anticipation of where the system is headed and what could happen next. Part of the difficulty is how to anticipate future trends without being trapped in predictive models, which may be too brittle, uncertain, and cumbersome to rely on given the variability of real fields of practice and the real consequences of failure.

ESCALATION

There is a fundamental relationship regarding tempo that drives how JCSs adapt (Woods & Patterson, 2000):

> **Escalation Principle**: The concept of escalation concerns a process— how situations move from canonical or textbook to non-routine to exceptional. In that process, escalation captures a relationship—as problems cascade, they produce an escalation of cognitive and coordinative demands that bring out the penalties of poor support for work.

The concept of escalation captures a dynamic relationship between the cascade of effects that follows from an event and the demands for work that escalate in response (Woods, 1994). An event triggers the evolution of multiple interrelated dynamics. The response following an anomaly during ascent of a space shuttle mission (Chapter 8, pp. 69-70, and Figure 7) illustrates the cascade and escalation of demands (the reader can use the case of "Being *Bumpable*" in Chapter 3 to practice charting these cascades themselves).

1. There is a cascade of effects in the monitored process.
A fault produces a time series of disturbances along lines of functional and physical coupling in the process (e.g., Abbott, 1990). These disturbances

produce a cascade of multiple changes in the data available about the state of the underlying process, for example, the avalanche of alarms following a fault in process control applications (Reiersen, Marshall, & Baker, 1988).

2. Demands for cognitive activity increase as the problem cascades.
More knowledge potentially needs to be brought to bear. There is more to monitor. There is a changing set of data to integrate into a coherent assessment. Candidate hypotheses need to be generated and evaluated. Assessments may need to be revised as new data come in. Actions to protect the integrity and safety of systems need to be identified, carried out, and monitored for success. Existing plans need to be modified or new plans formulated to cope with the consequences of anomalies. Contingencies need to be considered in this process. All these multiple threads challenge control of attention and require practitioners to juggle more tasks (increasing the risk of workload bottlenecks).

3. Demands for coordination increase as the problem cascades.
As the cognitive activities escalate, the demand for coordination across people, across groups, and across people and machines rises. Knowledge may reside in different people or different parts of the operational system. Specialized knowledge and expertise from other parties may need to be brought into the problem-solving process. Multiple parties may have to synchronize activities aimed at gaining information to aid diagnosis or to protect the monitored process. The trouble in the underlying process requires informing and updating others—those whose scope of responsibility may be affected by the anomaly, those who may be able to support recovery, or those who may be affected by the consequences the anomaly could or does produce.

4. The cascade and escalation is a dynamic process.
A variety of complicating factors can occur, which move situations beyond canonical, textbook forms (Woods et al., 1990). The concept of escalation captures this movement from canonical to non-routine to exceptional. The tempo of operations increases following the recognition of a triggering event and is synchronized around temporal landmarks, particularly those that represent irreversible decision points. The dynamics of escalation vary across situations. First, the cascade of effects may have different time courses. For example, an event may manifest itself immediately or may develop more slowly. Second, the nature of the responses by practitioners affects how the incident progresses—less appropriate or timely actions (or too quick a reaction in some cases) may sharpen difficulties, push the tempo in the future, or create new challenges. Different domains may have different escalation gradients depending on the kinds of complicating factors that occur, the rhythms of the process, and consequences that may follow from poor performance.

5. Interactions with support systems occur relative to demands.
Interactions with support systems (computer based or via other technologies) occur in the context of these escalating demands on knowledge and attention, monitoring and assessment, communication and response. In situations within the envelope of textbook competence, technological systems seem to integrate smoothly into work practices, so smoothly that seemingly little work is required for human roles. However, patterns of distribution and coordination of this work over people and machines grow more complex as situations cascade. Thus, the penalties for poor coordination between people and machines and for poor support for coordination across people emerge as the situation escalates demands.

The difficulties arise because interacting with the technological devices is a source of workload as well as a potential source of support. Interacting with devices or interacting with others through devices creates new burdens that can combine in ways that create bottlenecks. Practitioners are placed in an untenable situation if these new burdens occur at times when practitioners are busiest on critical tasks, if these new attentional demands occur when practitioners are already plagued by multiple voices competing for their attention, or if these new sources of data occur when practitioners are overwhelmed by too many channels spewing out too much competing data.

As active, responsible agents in the field of practice, practitioners adapt to workaround these bottlenecks in many ways—they eliminate or minimize communication and coordination with other agents, they tailor devices to reduce cognitive burdens, they adapt their strategies for carrying out tasks, they abandon some systems or modes when situations become more critical or higher tempo (*JCS-Foundations*, pp. 106-107). Woods et al. (1994) devote a chapter to examples of these workload bottlenecks and the ways that people tailor devices and work strategies to cope with forms of technology-induced complexity (cf., Sarter et al., 1997 for a summary on cockpit automation or Cook & Woods, 1996 for a study on operating room information technology).

Properties of JCSs, such as escalation, illustrate how typical questions about allocating tasks to either the human or the automation are fundamentally misguided and unproductive (*JCS-Foundations*, pp. 121-124). Instead of asking, "Who does what with this in isolation?" CSE asks, "How does the JCS adapt to changing demands?" For example, the advent of technology for unmanned aerial vehicles (UAVs) led most technologists and development managers to ask questions about how many human operators it takes to operate a UAV and to ask for assistance in reducing the ratio of people to UAVs. This is the wrong question and produces a wave of reactive ergonomics work that becomes quickly outdated by the next shift or leap in technology (yet another case of "dustbin human factors"). As a result, the search for design concepts becomes stuck in copying over old interfaces and roles into new situations and capabilities that leaves all stakeholders dissatisfied (in the UAV case, recruiting pilots to tele-operate UAVs).

The useful direction is to ask how the system of agents adapts to recognize and handle anomalies and opportunities. How is the system of agents able to escalate work to match cascading effects and tempo variations when situations challenge the envelope of textbook competence? Designing and testing to this target opens whole new territories for innovation such as, in the expanding technology for UAVs, how to help recognize and adapt to disruptions to the plan in progress, how to see if the automation's plan fits the situation that is unfolding, how to support integrating multiple feeds that capture different aspects of the remote situation of interest (Woods et al., 2004).

COUPLING

As one considers the cases and patterns in JCSs at work, note the role of coupling as a form of complexity. Coupling refers to the degree of interconnections between parts in a process to be controlled. Tighter coupling across parts of a process produces more complex disturbance chains following disruptions. Disturbances propagate more quickly and further through the monitored process. Tighter coupling intensifies all of the demands in anomaly response, including factors related to tempo of operations, coordination across parties, switching among multiple threads (control of attention), avoiding workload bottlenecks.[8]

Perhaps the most significant consequence of tighter coupling is the increase in potential side effects of any event that occurs, action that is taken, or plan that is modified. The development of techniques for goal-means analysis (Rasmussen & Lind, 1981; Rasmussen, 1986; Woods & Hollnagel, 1987; Vicente, 1999; Lind, 2003) are important because they provide a means to map couplings across functional and physical levels of description and capture how disrupting events produce cascades of effects (Woods, 1994).[9] By providing information on various forms of coupling, goal-means analyses contribute to tracking what side effects occur and how goals interact in specific situations.

---

[8] Note that organizational theorists and CSE both use the term coupling, but in different senses. CSE restricts coupling to refer to properties of the monitored process that affect how disturbances propagate which in turn creates and intensifies the demands of work. Organizational theorists use the label coupling also to refer to the spread of information and decision making across the management structure.

[9] It is important to avoid confusing means for end in analyses of JCSs at work. The abstraction hierarchy is one analysis mechanism and it is used to capture relationships across goals, functions, and physical mechanisms in a work domain (Rasmussen, 1986; Vicente, 1999). Describing a work domain in the form of an abstraction hierarchy can become an end in itself in cognitive task synthesis, rather than means to map couplings and help decompose the relationships across strategies, demands, and artifacts in work. The criterion to judge any cognitive task synthesis is always: does it provide a functional account how the behavior and strategies are adapted to the goals and constraints of the field of practice. Specific ways to analyze a work domain are useful to the degree they contribute to that end, for example, by helping to map contexts, side effects, interacting goals, and tradeoffs.

Tighter coupling means that disruptions and actions produce multiple effects which challenge coordination and resilience:

• Anomaly recognition and diagnostic search become more difficult since some of the follow-on disturbances can be physically or functionally distant from the original triggering event—"effects at a distance." As disrupting events produce multiple effects, the ability to discriminate red herrings from important "distant" indications will be more difficult (e.g., the scenario in Roth et al., 1992). This exacerbates the already difficult demand to focus in on the significant subset of data as context changes (Woods, 1995a).

• Increased coupling creates more situations where different goals interact and conflict. This means that there are more dilemmas or tradeoffs to be handled and resolved in practice (Woods et al., 1994).

• As coupling goes up, a single action will have multiple effects. Some of these will be the intended or main effects while others will be "side" effects. These side effects must be considered in replanning following disruptions to the plan in progress or assessing the actions of other agents, human or machine. Missing side effects in diagnosis, in planning, and in adapting plans in progress to cope with new events is a common vulnerability that contributes to failure in highly coupled systems (Woods & Shattuck, 2000). For example, when an event disrupts a plan in progress for a tightly coupled process, there are more reverberations to manage during replanning.

• Coupling expands the number of threads or lines of activity and reasoning that can be intertwined during work (directed attention and managing multiple threads). This increases the demand for effective control of attention in order to switch the focus of attention as conditions and priorities change (Woods, 1995b).

• When distant parts are coupled, coordination demands increase as practitioners in one role must know about other parts of the process, know more about the work carried out by other roles that monitor manage those parts, and know more about how activities in their scope of responsibility affects others and visa versa.

One of the innovations in support for JCSs at work (affordances) that has resulted from work in CSE is the need to develop side effect tracking displays for highly coupled systems (e.g., Woods & Hollnagel, 1987; Watts-Perotti & Woods, 1999).

PREMATURE NARROWING

When we step back and examine studies on anomaly response, one theme that recurs is the danger of premature narrowing. Through careful observation of the development of expertise and coordinative mechanisms in a JCS, one begins to notice the presence of strategies for avoiding this basic vulnerability (Cook et al., 2000). Findings have highlighted the danger of becoming stuck in one assessment and being unable to revise the assessment even as new evidence comes in or situations change (Woods et al., 1987). Studies of hypothesis generation in diagnostic reasoning found that the key is broadening the set of possible explanations to be considered (Gettys et al., 1987). Studies of professionals in information analysis found that premature narrowing was a basic vulnerability as analysts moved to new computer-based systems in order to cope with massive increases in data availability and were challenged by new tasks outside their home base of experience (e.g., Patterson et al., 2001; Elm et al., 2005).

All processes for anomaly response and information analysis must, within some horizon, funnel-in on key sources, on the basic unexpected finding to-be-explained, on the storyline that explains the unfolding events and evidence gathered, and on the critical action or plan revision that needs to be undertaken to accomplish goals in changing situations. The danger is a premature narrowing that misses or discounts evidence that would lead to revision. Experienced practitioners develop "broadening" checks that they combine with the normal funneling processes to reduce the risk of premature narrowing or closure. In effective performance, a JCS adjusts the sequence of funneling-in plus broadening checks to converge in a timely manner while remaining sensitive to the need to revise previous assessments. Using broadening checks is a balance between the need to be sensitive to the potential for misassessment and the need to accomplish work within time and resource bounds inherent in evolving situations or imposed to meet organizational goals. Notice that funneling-in plus broadening is another example of Neisser's perceptual cycle (*JCS-Foundations*, p. 20) and demonstrates how this concept is fundamental to analysis and synthesis of JCSs at work.

Criteria that assess broadening versus premature narrowing also are key in studies of collaboration. Studies of human-automation coordination find that poor collaborative architectures narrowed the range of data considered and hypotheses explored (Layton et al., 1994). Studies of human collaboration find that diversity across participants improves problem-solving performance (Smith et al., 1997; Hong & Page, 2002). Studies of error detection and correction point to the need for collaborative cross checks across the multiple practitioners involved in providing care (Patterson et al., 2004; or Fischer & Orasanu, 2000, for aviation). Thus, one part of the value of effective collaborative interconnections lies in how they can broaden focus, reduce mis-assessments, and support revision.

REFRAMING

The dynamic character of anomaly response has revealed the importance of revision. When anomaly response breaks down it is often associated with an inability to revise plans and assessments as new evidence arrives and as situations change. Failures to revise, as a basic vulnerability in JCSs, are a form of under-adaptation.

In the perceptual cycle (*JCS-Foundations*, p. 20), data noticed about the world trigger frames (including purposes) that account for the data and guide the search for additional data. At the same time, the current framing orients the observer to the world, changing what counts as data (e.g., the discussion of unexpected events). Both activities occur in parallel—data generating frames, and frames defining what counts as data. The revision of assessments is more than a simple adjustment to the current assessment, but rather the more difficult process of reframing (Klein et al., in press). For example, a review of research on problem detection found that failures of problem detection are not so much failures to detect an early indicator, but rather they are failures to re-conceive or redefine the situation (Klein et al., 2004).

Fixating on a situation assessment, thus missing or discounting new evidence is another example of the need for and difficulty of reframing. The basic defining characteristic of fixations is that the immediate problem-solving context has biased the problem-solver in some direction—framing. The data on successful and unsuccessful revision of erroneous situation assessments shows that reframing usually takes a person with a fresh point of view on the situation (Woods et al., 1987). For example, after the Three Mile Island accident (Kemeny, 1979), the nuclear industry took actions to avoid failures to revise assessments by adding a new role to the control room team with a different background and viewpoint (called the Shift Technical Advisor) and giving the operators new kinds of representation of the behavior of the plant relative to goals (new displays and procedures organized around safety functions).

As in this example, one can aid reframing by changing how different people in the system coordinate their roles to try to ensure a fresh point of view, i.e., one that is unbiased by the immediate context, or to ensure effective cross checks (Patterson et al., 2004). One can aid reframing by providing new forms of feedback on the behavior of the monitored process that captures events, future directions (what could happen next), views that integrate data around models of the work domain (Vicente, 1999), multiple views that provide contrasting perspectives on what data mean, and displays on goal achievement (Woods, 1995a).

DILEMMAS

... to see "very well that it was necessary to perish in order not to perish;
and to expose oneself to dangers of all kinds, in order to avoid all dangers.
Jesuit Relations (1656-57)

The rising complexities of practice create or exacerbate competing demands such as conflicting goals. Multiple, simultaneously active goals are the rule rather than the exception for virtually all domains in which expertise is involved. Practitioners must cope with the presence of multiple goals, shifting between them, weighing them, choosing to pursue some rather than others, abandoning one, embracing another. Many of the goals encountered in practice are implicit. Goals often conflict. Sometimes these conflicts are easily resolved in favor of one or another goal, sometimes they are not. Sometimes the conflicts are direct and irreducible, for example, when achieving one goal necessarily precludes achieving another one. But there are also intermediate situations, where several goals may be partially satisfied simultaneously. An adequate analysis of JCSs at work requires explicit description of the interacting goals, how they create tradeoffs and dilemmas, and how these are resolved in practice (Woods et al., 1994).

Some dilemmas are embedded in the very nature of technical work in that field of practice. These include technical dilemmas that arise from demands inherent in the process to be managed and controlled. For example, some situations demand that anesthesiologists attempt to keep blood pressure both high and low to meet different goals on the patient's cardiac system. For an anesthetized cardiac patient, a high blood pressure works to push blood through the coronary arteries and improve oxygen supply to the heart muscle. On the other hand, because increased blood pressure adds to cardiac work, a low blood pressure is desirable to reduce cardiac work. The appropriate blood pressure target adopted by the anesthesiologist depends in part on the practitioner's strategy, the nature of the patient, the kind of surgical procedure, circumstances within the case that may change (e.g., the risk of major bleeding), and the negotiations between different people on the operating room team (e.g., the surgeon who would like the blood pressure kept low to limit the blood loss at the surgical site).

Organizations constrain and pressure practice in ways that create or intensify dilemmas. For example, consider how increased pressure for efficiency creates more same-day surgeries, which may exacerbate dilemmas. In the daily routine, the goal of discovering important medical conditions with anesthetic implications before the day of surgery may drive the practitioner to seek more information about the patient. Each hint of a potentially problematic condition provides an incentive for further tests that incur costs (e.g., the dollar cost of the tests, the lost opportunity cost when a surgical procedure is cancelled and the operating room goes unused for that time, the collaborative costs of disgruntled surgeons). The goal of minimizing costs, in contrast, provides an incentive for the use of same-day surgery even though this constrains preoperative evaluations. Thus, the medical practitioners face a dilemma (Woods, 2006a). Some practitioners in some situations might not follow up hints about some aspect of the patient's history because to do so would impact the usual practices relative to throughput and economic goals. This is especially true since following up the hint will only rarely lead to important information, though the delay surely will interrupt the workflow and incur costs. Other practitioners will adopt a defensive stance and order tests for minor indications even though the yield is low, to be on the safe side. This generates increased costs and incurs the wrath of their surgical colleagues for the delays thus generated.

Organizational factors at the blunt end of systems shape the world in which practitioners work by influencing the means available for resolving dilemmas. In aviation, an aircraft is de-iced and then enters the queue for takeoff. After the aircraft has been de-iced, the effectiveness of the de-icing agent degrades with time. Delays in the queue may raise the risk of ice accumulation. However, leaving the queue to go back to an area where the plane can be de-iced again will cause additional delays, and in addition, the aircraft will have to re-enter the takeoff queue again. Thus, the organization of activities (where de-icing occurs relative to where queuing occurs in the system) can create conflicts that the practitioners must resolve because they are responsible at the sharp end of the system. Unfortunately, there have been aircraft crashes where, in hindsight, crews accepted delays of too great a duration and ice did contribute to a failed takeoff (Abbott, 1996). In general, balancing tradeoffs created by conflicts of safety versus production goals can prove difficult, as turned out to be the case in the Columbia space shuttle accident (CAIB, 2003; Woods, 2005b).

Some tradeoffs emerge in the nature of demands placed on the work of JCSs. In anomaly response, for example, we have seen how there is a trade-off with respect to when to commit to a course of action. Practitioners have to decide whether to take corrective action early in the course of an incident with limited information, to delay the response and wait for more data to come in, to search for additional findings, or to ponder additional alternative hypotheses (see Figure 11). In control of attention, there is the trade-off between being too easily interrupted by new signals or events and being too focused on the current priority, which affects the balance between the risk of vagabonding from thread to thread incoherently and the risk of failing to revise an assessment as the situation changes (Woods, 1995b).

Practitioners also trade off between following standard routines or adapting the routine to handle the particular situation they face in order to meet the intent behind the plan or procedure (cf., Shattuck & Woods, 2000). Do the standard rules apply to this particular situation after a disrupting event or when some additional complicating factor is present? This is the problem of coordinating the distant but global perspective of supervisors/management with the local and up-to-date perspective of the operator on the scene. Mis-balancing the trade-off risks the failures of under-adaptation—continuing to apply a plan that doesn't fit the situation at hand—or over-adaptation—ad hoc adaptations to disruptions that fail to take into account the broader goals and constraints. Supervisors and the larger organizational context must determine the latitude or flexibility they will give actors to adapt plans and procedures to local situations, given the potential for surprise in that field of activity. Supervision that establishes centralized control inhibits local actors' adaptations to variability, increasing the vulnerability to under-adaptation failures. At the other extreme is supervision that provides local actors with complete autonomy. In the latter case, the goals and constraints important in the remote supervisors' scope are disconnected from the activity and decision making of local actors. As a result, the response across multiple local actors may not be coordinated and synchronized properly, increasing vulnerability to over-adaptation failures. The systems concept of resilience (Hollnagel et al., 2006) suggests that organizations look for evidence of gaps between distant plans and the factors that challenge those plans. A large gap places demands on practitioners to adapt in order to resolve the dilemmas and conflicts.

Dilemmas also arise in advisory interactions across agents (Roth et al., 1987). A machine expert recommends a particular diagnosis or action, but what if your own evaluation is different? What constitutes enough evidence that the machine is wrong to justify disregarding the machine expert's evaluation and proceeding on your own evaluation of the situation?

The dilemmas may be resolved through conscious effort by specific teams, or practitioners can also simply apply standard routines without deliberating on the nature of the conflict. In either case, they may follow strategies that are robust (but still do not guarantee a successful outcome), strategies that are brittle (work well under some conditions but are vulnerable given other circumstances), or strategies that are very vulnerable to breakdown. Uncovering the dilemmas that occur in practice and how they are resolved through practice will reveal how behavior and strategies are adapted to constraints (and this provides a good tip for designing problems/scenarios to use in studies intended to discover how a JCS works).

Value of mapping goal—means relationships in the work domain lies in how the analysis reveals dilemmas and characterizes how multiple factors come together to actualize conflicts in particular situations.

OVER-SIMPLIFICATIONS

The cases discussed to this point also reveal a larger pattern about coping with complexity. When confronted by complexity, people have a tendency to simplify (JCS-Foundations, p. 82). This is a locally adaptive response wherever the complexity originates. The stories of adaptation in JCSs reveal simplifications as a coping response when a new artifact is misfit to demands such as control of attention, when an increase in coupling produces additional disturbance chains cascading from a disrupting event, or when management creates double binds for practitioners wrestling in specific situations with conflicts between simultaneously important but inconsistent goals.

The problem of modern systems is a rise in complexity that results from successful adaptation to the pressure to be "faster, better, cheaper"—or the Law of Stretched Systems (see NASA, 2000; Woods, 2005b; Woods, 2006a). Driven by demands for new performance levels and by pressures to reduce resources; plus, fuelled by new capabilities from the expanding powers of new technology for connectivity, for data pick-up/collection/ transmission, and for extending our presence into remote environments, fields of practice face new complexities. In addition, the rising tide of complexities to meet the pressure to be "faster, better, cheaper" undermines the viability of previous simplifications adopted to manage the demands and tradeoffs in work. CSE arose as a response to this growth in complexity (*JCS-Foundations*, pp. 3-5).

Feltovich and his colleagues have studied how people understand complex concepts (often highly trained practitioners such as cardiologists), and have found a set of over-simplification tendencies at work (cf. also Rasmussen, 1986, on simplifications as shortcuts in decision making). Table 4, taken from Feltovich, Spiro & Coulson (1997), captures some of the basic dimensions along which simplification and over-simplification can occur. First, note that these tactics no longer work well in the face of the complexities of modern work systems under "faster, better, cheaper" pressures. They become over-simplifications ill-suited to handle the situations, risks and uncertainties of modern systems, e.g., how the increased coupling or interactions between parts produce more complex cascades of effects to be tracked and controlled. Second, close examination of Table 4 reveals that simplifications along those dimensions are all essentially **narrowing** heuristics. When the work of JCSs is based on these heuristics, the system is vulnerable or exposed to failure by missing side effects of an action given the interconnections across parts, through poor synchronization of multiple lines of activity, by getting stuck on seeing one factor as the cause of outcomes when results actually emerge from the interactions and contributions of multiple factors.

Table 4. Over-simplifications (from Feltovich, Spiro & Coulson, 1997)

1. Discreteness/continuity. Do processes proceed in discernible steps, or are they unbreakable continua? Are attributes adequately describable by a small number of categories (e.g., dichotomous classifications like large/small), or is it necessary to recognize and utilize entire continuous dimensions (e.g., the full dimension of size) or large numbers of categorical distinctions?
2. Static/dynamic. Are the important aspects of a situation captured by a fixed "snapshot," or are the critical characteristics captured only by the changes from frame to frame? Are phenomena static and scalar or do they possess dynamic vectorial characteristics?
3. Sequentiality/simultaneity. Are processes occurring one at a time, or are multiple processes happening at the same time?
4. Mechanism/organism. Are effects traceable to simple and direct causal agents, or are they the product of more system-wide functions. Can important and accurate understandings be gained by understanding just parts of the system, or must the entire system be understood for even the parts to be understood well?
5. Separability/interactiveness. Do processes occur independently or with only weak interaction, or is there strong interaction and interdependence?
6. Universality/conditionality. Do principles hold in much the same way (without the need for substantial modification) across different situations, or is there great context-sensitivity in their applicability?
7. Homogeneity/heterogeneity. Are components or explanatory schemes uniform (or similar) across a system—or are they diverse?
8. Regularity/irregularity. Is a domain characterized by a high degree of routinizability across cases, or do cases differ considerably from each other even when commonly called by the same name? Are there strong elements of symmetry and repeatable patterns in concepts and phenomena, or is there a prevalence of asymmetry and absence of consistent pattern?
9. Linearity/nonlinearity. Are functional relationships linear or nonlinear (i.e., are relationships between input and output variables proportional or non-proportional)? Can a single line of explanation convey a concept or account for a phenomenon, or are multiple and overlapping lines of explanation required for adequate coverage?
10. Surface/deep. Are important elements for understanding and for guiding action delineated and apparent on the surface of a situation, or are they more covert, relational, abstracted?
11. Single/multiple. Do elements in a situation afford single (or just a few) interpretations, functional uses, categorizations, and so on, or do they afford many? Are multiple representations required (multiple schemas, analogies, case precedents, etc.)?

The functional synthesis of anomaly response points to factors that make it necessary to avoid the simplifications in Table 4. For example, as tempo and coupling increase, the danger of failures to revise increases. Thus, we begin to see a pattern where a routine, plan or algorithm is deployed correctly, but in the wrong situation—i.e., the actual the situation demands a different response (Mitroff's 1974 error of the third kind). This means design needs to provide new kinds of artifacts that promote observability of dynamic, multi-factor processes, new kinds of artifacts for coordination that support escalation of knowledge and expertise as a situation cascades, new forms of broadening checks and cross-checks that enhance revision of assessments and plans as situations change or disruptions occur (cf., *JCS-Foundations*, pp. 87-91).

These over-simplifications captured by Feltovich's research do not apply just to work at the sharp end of systems (Figure 2). Designers and managers at the blunt end of systems are quite vulnerable to over-simplification tendencies, as well, as they make investment and development decisions that affect the future of operations (Woods & Dekker, 2000). Over-simplification tendencies have dominated the investigation of how systems fail (the red herring of "human error" and linear causality in accident and root cause analysis (Hollnagel, 1993; Woods et al., 1994, chapter 6; Dekker, 2002; Hollnagel, 1998; Hollnagel et al., 2006).

As we begin to consider stories of work about miscoordination between automation and people in Chapter 10, note that a second story line is going on in parallel. The "second" story (Cook et al., 1998; Woods & Tinapple, 1999) is a story of how the blunt end is trapped in over-simplifications about how new technological powers affect the future of work and in over-simplifications about how people and machines coordinate as joint cognitive systems (see Roesler et al., 2001; Woods et al., 2004; Feltovich, Hoffman et al., 2004).

To summarize: Patterns in Multi-Threaded Work

Multi-threaded work challenges simplification heuristics. These heuristics become over-simplifications that expose the JCS to new vulnerabilities for failure. Describing, modeling and supporting multi-threaded work requires new ways to represent dynamic balances as in control of attention, new ways to manage cascades of effects from disrupting events, and new ways to re-conceptualize the problem to be solved or the key goal to be achieved.

Chapter 10

Automation Surprises

The operator's job is to make up for holes in designers' work.
(Rasmussen, 1981)[10]

Consider the following mishap where one can see the operation of many of the factors important in coordinating teams of people and automation.

Why Didn't You Stop the Automation ...?

On December 6[t], 1999, a Global Hawk UAV (unmanned aerial vehicle) "accelerated to an excessive taxi speed after a successful, full-stop landing. The air vehicle departed the paved surface and received extensive damage... when the nose gear collapsed" ($5.3 million). How did this automated vehicle end up nose down in the desert 150 yards off the runway (see Figure 12)? The mishap began with an in-flight failure and "the air vehicle transitioned to a preplanned contingency route and returned to base. After the UAV landed and stopped on the runway, the command and control officer commanded the vehicle to taxi." Up to this point everything had gone according to the textbook. Following the human supervisor's instruction to proceed with taxi plan, "the air vehicle accelerated in an attempt to attain the preprogrammed commanded ground speed of 155 knots." It proceeded to the next programmed waypoint and then executed a turn to the left to reach the following waypoint specified in the plan. Needless to say the aircraft could not execute the turn while accelerating to such a high speed and skidded off the runway into the desert.

How did the UAV get instructions to taxi at 155 knots? Why didn't the human supervisors recognize the unreasonable taxi speed and stop the automation before it began the turn? The accident investigation revealed that a software problem with some of the contingency plans led the automation, when instructed to continue with the taxi plan, to look up and change the speed to a climb speed (155 knots) rather than a taxi speed (about 6 knots). The software problem resulted from hidden dependencies (see Rae et al., 2002) introduced during software updates (and related to the catalog of known software bugs that had been cumulating for redress) and from limits on software testing (software testing did not include taxi phases and mission planning was unable to analyze in sufficient depth UAV behavior over all contingencies as the mission planning process was already very long and effortful).

---

[10] From a talk by Jens Rasmussen at the IEEE Standards Workshop on Human Factors and Nuclear Safety, Myrtle Beach SC, August 30—September 4, 1981.

Given that software problems resulted in an erroneous ground speed in the taxi phase, why didn't human supervision detect the problem and respond to prevent the runway excursion? The various people monitoring the vehicle were focused on checking fuel imbalance anomalies. Their first indication of something amiss was when the vehicle exceeded normal taxi speed, at which time it was already virtually too late to stop the vehicle in time (update rates on the data displays were quite slow). Still the personnel recognized the excessive speed and issued a command to abort (20 seconds after taxi began). However, the vehicle did not respond to the abort command (or to a second abort command 15 seconds later), and eventually the supervisors commanded an emergency fuel shutdown to the engines (after 40 seconds), but the aircraft had already come to rest in the desert at this point. Later it was determined that a loss of communication had occurred; brief network dropouts were a relatively common experience. Nevertheless, the accident board concluded that, while excessive speed was the cause, a "contributing factor … was a breakdown in supervision."

Since development of the UAV focused on achieving higher autonomy, little effort went into understanding the operator's new role as supervisor and no integrated visualizations were provided (only "elemental" displays that showed each aircraft parameter individually and each mode status for the automation were available). The supervisor's main recourse was to abort the plan and bring the aircraft to a stop (typically, provisions exist for the supervisor to takeover manual control). The result was that the supervisors could not see what the automation was set up to do, they could not see that it was about to carry out a plan that did not fit the actual context, and they had limited means to re-direct the automated system other than switching to a form of manual control. Since the system was "automated," the people could be re-tasked to carry out other duties related to the larger cycle of planning missions, instructing automation, and improving the time cycle; hence, they were monitoring other issues that could affect turnaround (fuel anomalies).

Global Hawk UAV, 98-2003, 19991206, FSPM 1201A

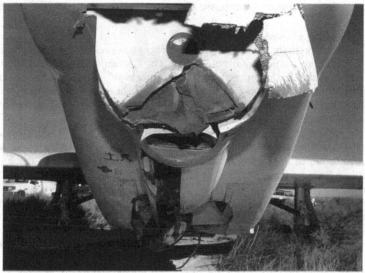

Photo #3   **SENSOR DAMAGE**

Global Hawk UAV, 98-2003, 19991206, FSPM 1201A

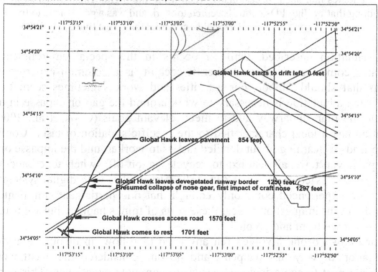

IMPACT SITE

Figure 12. Global Hawk UAV mishap (From accident report; U.S. Air Force, 1999). Top. Damage to UAV. Bottom. Diagram of inability of UAV to execute a turn while taxing due to acceleration to an excessive speed.

Let us examine a different kind of case where practitioners carrying out workaday tasks confronted an impasse due to a glitch in computerized systems.

Working around an impasse.
During preflight at the gate, the crew of a glass cockpit aircraft is responding to an ATC (air traffic control) clearance. The Captain intended to set 9000 ft. in the *alt* (altitude) window as part of the process of setting up aircraft for the departure (a rotary dial to give an input to the flight computers). However, as he rotated the control, the electronic display only rotated to 999 and no higher—the lead digit had frozen at zero.

Captain turned it, turned it, turned it, stopped, looked at it. This cycle was repeated at least twice, perhaps four times; the co-pilot said let me try it, and he turned it, turned it, stopped, looked at it, repeating this cycle at least twice. Then both sat there staring at the recalcitrant device—Why won't it let me do this? What should we do next?

An FAA inspector was present giving the crew a check ride, and they were very embarrassed at being stuck. The crew was about to call maintenance and were thinking about how to handle the resulting delay.

At this point the FAA inspector blurted out, "Why don't you cycle the FDs [Flight Directors, an automated subsystem that computes and shows how the pilot should fly the aircraft];" the Captain reached up and did this right away, then he tried the knob again and successfully set the desired value.

Later in the flight the crew and inspector discussed that glitch and other kinds of glitches that occur occasionally when using the digital flight systems; included noting that cycling FDs control seemed to work and was a good tactic to try.
Personal communication, Test Pilot, July 1995

The situation described, while it occurs in the special environment of an automated commercial transport aircraft cockpit, is a common experience. The systems that should support our activities and work sometimes seem to thwart them. Some adaptation is necessary to work around the gap or impasse in order to continue the plan underway and to meet relevant goals (Cook et al., 2000). The adaptation has a local character tied to the specific situation or task. Confronting the gap leads to learning about detailed tactics to workaround the impasse or bridge the gap—haven't we all learned to recycle an on-off switch to escape when a computerized device seems stuck? Inexplicable glitches with digital systems in the cockpit are experienced commonly enough that one participant commented later during the flight in question, "Yea, these sorts of things happen, but we usually get it to do what we want and go on."

Notice how no one involved felt any need to follow up or pursue any further investigation of why this happens, and no one considered the event worthy of making a report to some incident system to cumulate experience. This eliminates any opportunity to consider potential side effects of the action or how it might interact with or even create other problems. In one study of poorly designed computerized devices for anesthesiologists in the operating room, the observers found that physicians recycled an on-off switch thinking they were silencing nuisance alarms when they were actually changing between hidden modes which had other consequences (Cook et al., 1991).

What do these local workarounds mean? Are they clever adaptations? Are they a shortcut or compromise of proper procedures? This particular story of a workaround has a special flavor because of the presence of a representative of the regulatory authority. Did the pilots skip trying normal workarounds thinking the inspector would disapprove even though the call to maintenance would result in a departure delay and other consequences associated with a formal repair request? Or can we imagine a slightly different organizational context where the pilots are uncomfortable with a workaround, but are working under substantial production pressure? In this case, would the presence of the inspector provide them the additional justification needed to call maintenance despite the anticipation of management disapproval given the cost of a delay?

How are these adaptations discovered, learned, and transmitted among the community of practitioners? How does the organization react—curious to learn about shortfalls in their system? Is there any critical examination and learning from these adaptations to find where they are valuable, what the boundary conditions or side effects of these workarounds are, or to recognize which create larger risks because of otherwise unappreciated interconnections? Or does management threaten sanctions if procedures are violated? Note how the latter response creates a double bind for the practitioner—not working around the impasse is disapproved because of the delays and costs that would result, but working around the impasse also is disapproved and exposes the practitioner to disciplinary actions by management (Woods et al., 1994).

THE SUBSTITUTION MYTH

Interface: An arbitrary line of demarcation set up in order to apportion the blame for malfunctions.
Kelly-Bootle, 1995, p. 101

It's not cooperation, if either you do it all or I do it all.
First Law of Cooperative Systems, Woods, 2002

Cartoons often cut to the heart of a culture's underlying beliefs. In one cartoon a person is sitting in front of a computer system obviously experiencing difficulties getting the system to do what the user wants. Suddenly a genie appears and promises the user one wish. The user promptly says, "I wish you would make the problems with this computer system disappear" and "Poof!" the **user** disappears.

The cartoon and the Global Hawk mishap capture the basic cultural belief that emphasizes a gulf between people and information processing machines. The core assumption is that indications of trouble (incidents, workarounds, accidents) are evidence that people are an erratic unreliable component. Problems indicate the need to accelerate development of machines that can *substitute* for people—what we called over 20 years ago, "just a little more technology will be enough, this time."

This assumption (*JCS-Foundations*, pp. 101-102), while strong and persistent, is wrong. The gap between what researchers observe in terms of the actual effects that follow each wave of new technology and the claims of advocates for investment in and adoption of that technology reveals the problem (Woods & Dekker, 2000). Twenty years apart, the following statements use virtually the same words to describe the consequences of developing new computer-based intelligent or agent systems as if it were only a simple substitution of one agent for another.

> **1984:** ... one of the big problems is the tendency for the machine to dominate the human ... consequently an experienced integrated circuit designer is forced into an unfortunate choice: let the machine do all the work or do all the work himself. If he lets the machine do it, the machine will tell him to keep out of things, that it is doing the whole job. But when the machine ends up with five wires undone, the engineer is supposed to fix it. He does not know why the program placed what it did or why the remainder could not be handled. He must rethink the entire problem from the beginning.
>
> Finegold, 1984 (quoted in Woods, 1986a)

> **2005:** My concern is that the human is now being left out of the equation. So the systems we're designing are being thought of as stand-alone systems. Yet if you ask these people about how they'll be employed in real-life situations, for example at border crossings, what they say is that the machine will be used to filter out difficult cases, and those difficult cases where the machine can't decide whether or not this is the person who they claim to be, will be handed over to a human operator, without any consideration of how well the human will do this task. In fact we know the human will find that a rather difficult task. No one has ever considered what happens when the human is only given the difficult tasks to deal with. So in other words, the machine system and the human need to be considered together as a single entity.
>
> Richard Kemp, 2005

Surprises about Automation

The idea that new technology can be introduced as a simple substitution of machines for people—preserving the system though improving the results—is a persistent over-simplification fallacy at the blunt end of systems (Table 4). Observing the introduction of new technology and systems into an ongoing field of practice—the reverberations and adaptations that follow—reveal patterns that are quite different from the substitution myth (Flores et al., 1988; Carroll et al., 1991; Woods & Dekker, 2000). Observations reveal that change represents new ways of doing things; i.e., it does not preserve the old ways with the simple substitution of one medium for another (e.g., paper for computer-based) or one agent for another (automated system for human operator). Adding or expanding the machine's role changes the JCS, changing roles, coordination, and vulnerabilities. Understanding and guiding the process technology change comes only with the shift to a joint systems view that emphasizes the interactions between people, technology, and work (Hollnagel & Woods, 1983; JCS-Foundations, Chapter 1).

The story of the Global Hawk mishap reveals how technology change (UAV capabilities) transforms work as operators needed to monitor and re-direct the automated capabilities. The Global Hawk mishap is a story of how autonomous systems are brittle at the boundaries of their capabilities and the kinds of factors that arise that challenge the model of operations assumed by the builders of the automation—the kinds of inevitable "holes" in designers' work that operators must compensate for if systems are to be sufficiently resilient (Cook et al., 2000). The story of the accident also reveals that developers trapped in the substitution myth missed these predictable effects on the JCS, and failed to support the new roles and judgments with new forms of feedback and new tools for coordination.

Our fascination with the possibilities afforded by new technological powers often obscures the fact that new computerized and automated devices also create new burdens and complexities for the individuals and teams of practitioners responsible for operating, troubleshooting, and managing high-consequence systems. Additionally, the introduction of new technology is part of processes of change that transform work in these fields of practice. Table 5 summarizes some of the gaps between the apparent simplicity of the developer's eye view and the complexities experienced by practitioners.

What happens when new autonomous systems are designed and introduced as if they are independent agents without considering the JCS? The best-studied case is the introduction of automation into the commercial transport aircraft cockpit (Billings, 1997). When researchers applied natural history and staged world techniques as discussed in Chapter 5 to understand the reverberations of this case of technology change, they found stories of coping with complexity. Pilots and instructors described and revealed in their behavior the clumsiness and complexity of many modern cockpit systems. They described aspects of cockpit automation that were strong but sometimes silent and difficult to direct when the tempo of operations was high. New challenges were imposed by the tools that were supposed to serve them and to provide "added functionality."

The users' perspective on the automated systems is best expressed by the questions they ask each other and use with researchers when describing incidents (extended by Woods & Sarter, 2000 from Wiener, 1989):

- What is it doing now?
- What will it do next?
- How did I get into this mode?
- Why did it do this?
- Stop interrupting me while I am busy.
- I know there is some way to get it to do what I want.
- How do I stop this machine from doing this?
- Unless you stare at it, changes can creep in.

These questions and statements illustrate why one observer of human-computer interaction defined the term "agent" as: "A computer program whose user interface is so obscure that the user must think of it as a quirky, but powerful, person ..." (Lanir, 1995, p. 68).

Questions and statements such as the above indicate coordination breakdowns between people and automated systems that have been called *automation surprises* (Sarter, Woods & Billings, 1997). In these situations, crews are surprised by actions taken (or not taken) by the automated system just as the supervisors of the Global Hawk aircraft were surprised when they saw the vehicle accelerating to such a high taxi speed.

Table 5. Designer's eye view of apparent benefits of new automation contrasted with the real experience of operational personnel.

Putative benefit	Real complexity
Better results, same system (substitution)	Transforms practice and the roles/coordination of people
Frees up resources: 1.offloads work	Creates new kinds of work, often at the wrong times
Frees up resources: 2. focus user attention on the right information/answer	More threads to track; makes it harder to control attention
Less knowledge	New knowledge/skill demands
Autonomous machine	Team play with people is critical to success
Same feedback	New levels and types of feedback are needed to support people's new roles
Generic flexibility	Explosion of features, options and modes create new demands, types of errors, and paths toward failure
Reduce human error	Both machines and people are limited resource and fallible; new problems associated with coordination breakdowns between human and machine agents

Automation surprises begin with miscommunication and misassessments between the automation and users which lead to a gap between the user's understanding of what the automated systems are set up to do, what they are doing, and what they are going to do (this discussion is based on Woods & Sarter, 2000). In the Global Hawk case, the aircraft actually stops following touchdown so that the human supervisor can review the situation before instructing the automation to continue with the plan. However, there was no means for the supervisors to see the plan to be followed and check its appropriateness for the current situation.

The initial trigger for such a mismatch can arise from several sources, for example, erroneous inputs such as mode errors or indirect mode changes where the system autonomously changes its status and behavior based on its interpretation of user inputs, its internal logic and sensed environmental conditions (Sarter & Woods, 1995). In the Global Hawk mishap the trigger was hidden dependencies across software modules (for a similar case, see the Ariane 501 launch explosion, June 4, 1996).

Once the gap begins, the supervisors are surprised later when the aircraft's behavior does not match human monitors' expectations. This is where questions like, "Why won't it do what I want?" or "How did I get into this mode?" arise. It seems that the practitioners generally do not notice their misassessment from displays of data about the state or activities of the automated systems. The mis-assessment is detected, and thus the point of surprise is reached, in most cases based on observations of unexpected and sometimes undesirable aircraft behavior as in the Global Hawk case. Once the people have detected the gap between expected and actual aircraft behavior, they can begin to respond to or recover from the situation. The problem is that this detection generally occurs when the aircraft behaves in an unexpected manner—flying past the top of descent point without initiating the descent, flying through a target altitude without leveling off, or trying to accelerate to high speed while making turns on taxi. If the detection of a problem is based on actual aircraft behavior, it may not leave a sufficient recovery interval before an undesired result occurs. Unfortunately, there have been accidents where the misunderstanding persisted too long to avoid disaster (cf., Billings, 1997).

Woods & Sarter (2000) found the potential for automation surprises is greatest when three factors converge:

1. Automated systems act on their own without immediately preceding directions from their human partner
2. Gaps occur in users' models of how their machine partners work in different situations
3. Weak feedback is given about the activities and future behavior of the agent relative to the state of the world.

The surprising result is that new levels of autonomy need to be balanced with new forms of feedback or observability on the current and future behavior of the automated system. Observabilty is *feedback that provides insight into a process* and refers to the work needed to extract meaning from available data. To make automated systems behavior observable, displays need to meet three criteria:

1. Transition-oriented to capture and display events and sequences,
2. Future-oriented to reveal what should happen next and when,

3.   Pattern-based to support scanning at a glance to quickly recognize unexpected or abnormal conditions.

Again we see how observability is distinct from data availability, which refers to the mere presence of data in some form in some location. The critical test of observability is when displays help practitioners to notice more than what they were specifically looking for or expecting.

Note that none of these criteria were met in the Global Hawk case: while all of the necessary data were available somewhere in the information system, it had not been organized and represented to capture patterns as the people had to read and integrate each individual piece of data to make an overall assessment. No information was presented about the plan; this would include events, sequences, triggers, contingencies and activities the automation would perform. Displays only showed current settings and states with no support to help the person see what the automation would do next.

Studies of the impact of new automation revealed more than operational complexity; they also uncovered adaptations to cope with complexity in order to avoid coordination breakdowns with the capable but opaque and difficult to direct automated systems.   Training departments, line organizations, and individuals developed ways to get the job done successfully (through policies, procedures, team strategies, individual tactics and tricks) **despite** the clumsiness of some automated systems for some situations. Some of these are simply cautionary notes to practitioners reminding them to "be careful, it can burn you." Some are workarounds embodied in procedures. Some are strategies for teamwork. Many are ways organizations adopted to restrict the use of portions of the suite of automation in general or in particularly difficult situations.

For example, one place these adaptations could be seen was in examining changes to flight crew training programs. However, training managers expressed concern over being trapped in a decreasing window of time and resources, while the new cockpit systems required mastery of new knowledge and skills:

• "They're building a system that takes more time than we have for training people."
• "There is more to know—how it works, but especially how to work the system in different situations."
• "The most important thing to learn is when to click it off."
• "We need more chances to explore how it works and how to use it."
• "Well, we don't use those features or capabilities."
• "We've handled that problem with a policy."
• "We are forced to rely on recipe training much more than anyone likes."
• "We teach them [a certain number of] basic modes in training, they learn the rest of the system on the line."

Economic and competitive factors produced great pressure to reduce the training investment (e.g., shrink the training footprint or match a competitor's training footprint). Plus, management believed that greater investments in automation promised lower expenditures on developing human expertise. However, the data consistently show that the impact of new levels and types of automation is new knowledge requirements for people in the system as their role changes increasingly to that of a manager and anomaly handler (e.g., Sarter et al., 1997).

Despite deficiencies in the design of the automation as a team player, relatively few bad consequences occurred because of human expertise and adaptation (Woods et al., 1994, chapter 5). However, the list of accidents due to coordination breakdowns between people and automated systems appears to be growing. Besides the Global Hawk accident, there have been a series of mishaps (for summaries of the aviation cases see Abbott, 1996; Billings, 1997):

- Where the automation was driving the process under control (other salient examples are the Airbus A330 Test Flight, 4-2-95; the Ariane 501 explosion, 6-4-96; Mars Climate Orbiter science mission, 2-23-99; and Mars Polar Lander science mission, 12-3-99).
- Where people and automation fought for control of the process (e.g., Habsheim A320 aircraft crash, 6-26-88; Nagoya A300-600 aircraft crash, 4-26-94; Orly A310 aircraft incident, 9-24-94 and similar sequence at Moscow, 2-11-91); Warsaw A320 accident, 9-14-93).
- Where the automation to human operator handoff broke down (e.g., China Air aircraft near miss, 2-19-85 and other high altitude upsets; Roselawn ATR-72 aircraft crash, 10-31-94).
- Where the automation misguided the operator (e.g., Solar and Heliospheric Observatory science mission incident, 6-30-98; Bangalore A320 aircraft crash, 2-14-90).
- Where people and automation mis-coordinated plans (e.g., Therac-25 radiation overdoes in 1985 and 1986—see Leveson & Turner, 1993; Strasbourg, or Mt Saint-Odile, A320 aircraft crash, 1-20-92; Cali B757 aircraft crash, 12-20-95; and many medication mis-administrations through infusion devices, see Cook et al., 1992 for several near misses; Lin et al., 1998).

How do developers and technologists react when confronted with these mishaps? Woods & Sarter (2000) captured typical responses when developers in the aviation industry were confronted with evidence of a breakdown in the coordination between people and automation. Very similar responses have been collected from developers and management following human-automation mishaps in health care such as with infusion devices (see Cook et al., 1998; Doyle & Vicente, 2001 including the reply by C. H. McLeskey plus the debate in *Anesthesia Patient Safety Newsletter*, 15, 3, 36-39, 2000):

1. The hardware/software system "performed as designed."
2. The incident was due to "erratic" human behavior (variations: "You can't predict all of things people will do;" "Even skilled people sometimes do odd things;" "We're helping to train the users so they don't do this again").
3. The hardware/software system is "effective in general and logical to us; some other people, organizations, or countries just don't understand it" (e.g., those who are too old, too computer phobic, or too set in old ways).

4. "We would have gone further but..." We were constrained by what the customer insisted on, compatibility with the previous design, supplier's standard designs, cost control, time pressure, regulations...
5. Other parts of the industry "haven't kept up" with the advanced capabilities of our systems.
6. "Our next release/device/system is better (not that there is anything wrong with the last one);" "We're better than we used to be;" numerous improvements to the device have been made over the years; the device's overall performance record has been excellent; an upgrade is available.

While these rationalizations do reflect real and serious pressures and constraints in the design world, they reveal the depth of the substitution myth of people versus computers. This assumption forces the debate about what to learn from incidents into a line dividing reactions into either due to "human error" (which justifies more autonomous technology) or due to "over-automation" (which can only offer a pause or retreat from advances in technology). We can only escape from going round and round in this cycle—clumsy use of technology, new complexities, operational adaptation, occasional failures leading to more clumsy use of technology—by shifting to a systems perspective. When we use the joint cognitive system as the base unit of analysis, and examine how technological and organizational change transforms work in JCSs, we find a different and direct diagnosis—increases in the autonomy and authority of machines require an investment in more sophisticated forms of coordination across human and machine agents. Otherwise, new episodes of technology change will produce repeats of the results of past episodes: automation surprises will occur as a symptom of poor design for coordination in the face of unanticipated perturbations.

BRITTLENESS

Technology change is often justified on the basis of reducing the human role given the presumptions of the substitution myth. A diagnosis based on the JCS perspective shifts the question to how to overcome brittleness and enhance the adaptive power or resilience of the larger human-automation ensemble.

In complex settings where new mobile robotic systems are being deployed, difficulties cascade and demands escalate (Woods & Patterson, 2000). This will challenge robotic systems ability to compensate and will demand coordination between people and robots. Inevitably, automation capabilities will exhibit brittleness as situations develop beyond their boundary conditions (e.g., Roth et al., 1987; Guerlain et al., 1996; Smith et al., 1997). How will human team members recognize the approach to brittle boundaries and intervene effectively (e.g., bumpless transfer of control)? Inevitably, autonomous resources will be lost or will fail. How will the team dynamically reconfigure or gracefully degrade as assets are lost? One of many examples in human-robot interaction that tests coordination across these human robot handler and robot capabilities is judgments of traversability or climbability in context. If the robot provides valued access to remote environments, robot handlers want to avoid situations where the robot may get stuck or waste limited energy resources in circumstances where terrain conditions that challenge the robotic platform's capabilities. How will the robot handler role recognize or anticipate such conditions and how will they re-direct the robot's plan or adapt its behavior in these conditions? Woods et al. (2004) summarized the situation for the case of mobile robots based on Robin Murphy's operational studies of using robots in search and rescue (Murphy, 2004; Casper & Murphy, 2003; Burke et al., 2004):

> *(Robin)* **Murphy's Law**: Any deployment of robotic systems will fall short of the target level of autonomy, creating or exacerbating a shortfall in mechanisms for coordination with human stakeholders.

As robotic system developers strive to achieve a certain level of autonomy, in general, they underestimate the need for coordination with human stakeholders. Deployment into a field and context will leave the robotic system short of the design target level of autonomy, without sufficient provision for human stakeholders' involvement in handling the situation with or through the robotic system.

The research and experience base shows that, as autonomy and authority of automata increase, the demands for more sophisticated forms of coordination increase as well (Christoffersen & Woods, 2002). This is in stark contrast to the common beliefs that expanding the capabilities of automata reduces human roles in the system and reduces the need for coordination with those people. The difficulty is that, in envisioning future technological capability and future operational systems, it is easy to underestimate the demands for coordination. And, as Murphy (2004) has noted, development always will be aiming higher than it can currently reach in terms of autonomous capability so that pressures to deploy in the shorter run (from new operational pressures/demands) will add to the shortfall in support for coordination.

Coordination across agents is much more than considering what one agent alone will do (the traditional topic of allocating functions between people and machines; *JCS-Foundations*, pp. 121-123) or debating the degree of substitution between human and machine agents (usually cast in terms of "levels" of automation, e.g., Sheridan, 1992). These commonplace efforts only capture how much machines substitute for people and miss all of the functions needed for high degrees of coordination in work (Klein, Feltovich et al., 2005). In addition, such attempts ignore issues about responsibility in work (see pp. 151-156; Billings, 1997).

In contrast to the stories that began this chapter, collaborative problem solving occurs when the agents coordinate activity in the process of solving the problem. Designing support for coordination recognizes that machine agents are fundamentally brittle and literal minded—finite cognitive systems developed within finite development projects (the Bounded Rationality Syllogism, p. 2; Law of Demands, p. 19). Coordinating activities across agents given the demands of work, not simply enhancing capabilities or expertise of individual agents, creates success and resilience (Woods, 2002; 2006b).

MANAGING WORKLOAD IN TIME

In complex systems activity ebbs and flows, with periods of lower activity and more self-paced tasks interspersed with busy, high-tempo, externally paced operations where task performance is more critical. Introducing new technology into ongoing fields of human activity changes the distribution of workload over tasks, agents, and the dynamics of situations. A new artifact represents new levels and types of operator workload such as device setup and initialization, configuration control, operating sequences, interface management, new attentional demands, the need to track automated device state and performance, new communication or coordination tasks, and new knowledge requirements.

In the substitution myth, technology development simply shifts workload from the human to the machine so that overall human loads drop, allowing people to be removed from that part of the overall process. But the critical property for design of the JCS is how technology change impacts on low workload/tempo and high workload/tempo periods, and especially how it impacts the practitioner's ability to manage workload in order to avoid or cope with bottlenecks. One of the critical differences between clumsy and skillful use of the technological possibilities is the difference between undermining and supporting these processes of workload management.

To see workload management processes, one can go back to the earlier case of the intensive care unit as captured in Chapter 3 through the story "Being *Bumpable*." The cognitive task synthesis there reveals how practitioners adapt to workload bottlenecks and to the potential to be trapped in workload bottlenecks. In managing workload there are only four coping strategies: (1) shed load, (2) do all components but do each less thoroughly, thereby, consuming less resources, (3) shift work in time to lower workload periods, (4) recruit more resources.

The first two are tactical responses to emerging bottlenecks. When facing a bottleneck, one can prioritize across tasks and activities, dropping out any but the essential ones. But shedding load is a narrowing process that has vulnerabilities should the priority be misplaced. Similarly, in a bottleneck, one can take energy/time from every task, though now each task becomes more fragile. The paradox in tactical responses to workload bottlenecks is that, if there are tasks which really are lower priority, why not always drop them, or if task performance does not degrade when one cuts the normal investment of energy/time, why would one ever make the larger investment?

The third and fourth adaptive responses are strategic and depend on anticipation or knowledge of potential upcoming bottleneck points. With anticipation one can recruit more resources such as extra staff, special equipment, additional expertise, or additional time. But note that this strategy consumes organizational resources that are always under some degree of pressure (the Law of Stretched Systems, p. 18). More directly under the control of practitioners is the other strategic response—shift workload in time. For example, Cook & Woods (1996) observed that anesthesiology residents performed extra work during the set up of the operating room before the patient entered, in order to avoid potential workload bottlenecks should certain contingencies arise later. When anesthesiologists needed some type of capacity to respond to physiological changes in a patient, they rarely had the time or attention resources available to carry out the tasks required to create the needed capability. Hence, expertise in anesthesiology consists, in part, of being able to anticipate potential bottlenecks and high tempo periods and to invest in certain tasks which prepare the practitioner should the need arise (an orientation to intervene).

A syndrome which Wiener (1989) originally termed, "clumsy automation," occurs when the benefits of the new technology accrue during workload troughs, and the costs or burdens imposed by the technology occur during periods of peak workload, high-criticality, or high-tempo operations. Despite the fact that these systems are often justified on the grounds that they would help offload work from harried practitioners, we find that the new systems impose additional tasks, force the user to adopt new strategies, and require more knowledge and more communication at the very times when the practitioners are most in need of true assistance. This creates opportunities for new kinds of vulnerabilities to failure that did not exist in simpler systems.

Clumsy automation was first noted in the interaction between pilots and flight management computers (FMCs) in commercial aviation (Billings, 1997; Sarter & Woods, 1995). Under low-tempo operations, pilots communicate instructions to the FMCs which then "fly" the aircraft. Communication between pilot and FMC occurs through a multifunction display and keyboard. Instructing the computers consists of a relatively effortful process involving a variety of keystrokes on potentially several different display pages and a variety of cognitive activities such as recalling the proper syntax or where data are located in the virtual display page architecture. Pilots speak of this activity as "programming the FMC."

Cockpit automation is flexible also in the sense that it provides many functions and options for carrying out a given flight task under different circumstances. For example, the FMC provides at least five different mechanisms at different levels of automation for changing altitude. This customizability is construed normally as a benefit that allows the pilot to select the mode or option best suited to a particular flight situation (e.g., time and speed constraints). However, it also creates demands for new knowledge and new judgments. For example, pilots must know about the functions of the different modes, how to coordinate which mode to use when, and how to "bumplessly" switch from one mode or level of automation to another. In other words, the supervisor of automated resources must not only know something about how the system works, but also know how to work the system. Monitoring and attentional demands are also created as the pilots must keep track of which mode is active and how each active or armed mode is set up to fly the aircraft.

For example, studies showed that it was relatively easy for pilots to lose track of the automated systems' behavior when circumstances directed their attention to other activities (Sarter & Woods, 1995). Pilots would miss mode changes that occurred without direct pilot intervention during the transitions between phases of flight or during the high-workload descent and approach phases in busy airspace. These difficulties with mode awareness reduced pilots' ability to stay "ahead" of the aircraft—i.e., anticipate upcoming situations, tasks, and contingencies and anticipate the behavior of the automation in these situations.

Pilots developed strategies to cope with the clumsiness and complexities. For example, data indicate that pilots tend to become proficient or maintain their proficiency on a subset of modes or options. As a result, they try to manage the system within these stereotypical responses or paths, underutilizing system functionality. The results also showed that pilots tended to abandon the flexible but complex modes of automation and switch to less automated, more direct means of flight control, when the pace of operations increased (e.g., in crowded terminal areas where the frequency of changes in instructions from air traffic control increase). Note that pilots speak of this tactic as "escaping" from the automation.

Another study of clumsy automation examined the impact of operating room information systems and revealed some other ways that new technology creates unintended complexities and provokes practitioner coping strategies (Cook & Woods, 1996). In this case a new operating room patient monitoring system was studied in the context of cardiac anesthesia. This and other similar systems integrate what was previously a set of individual devices, each of which displayed and controlled a single sensor system, into a single CRT display with multiple windows and a large space of menu-based options for maneuvering in the space of possible displays, options, and special features. The study consisted of observing how the physicians learned to use the new technology as it entered the workplace.

By integrating a diverse set of data and patient monitoring functions into one computer-based information system, designers could offer users a great deal of customizability and options for the display of data. Several different windows could be called depending on how the users preferred to see the data. However, these flexibilities all created the need for the physician to interact with the information system—the physicians had to direct attention to the display and menu system and recall knowledge about the system. Furthermore, the computer keyhole created new interface management tasks by forcing serial access to highly inter-related data and by creating the need to periodically declutter displays to avoid obscuring data channels that should be monitored for possible new events.

The problem occurs because of a fundamental relationship: the greater the trouble in the underlying system or the higher the tempo of operations, the greater the information processing activities required to cope with the trouble or pace of activities. For example, demands for monitoring, attentional control, information, and communication among team members (including human-machine communication) all tend to go up with the tempo and criticality of operations. This means that the burden of interacting with the display system tends to be concentrated at the very times when the practitioner can least afford new tasks, new memory demands, or diversions of his or her attention away from patient state to the interface per se.

The physicians tailored both the system and their own strategies to cope with this bottleneck. In particular, they were observed to constrain the display of data into a fixed, spatially dedicated default organization rather than exploit device flexibility. They forced scheduling of device interaction to low criticality self-paced periods to try to minimize any need for interaction at high workload periods. They developed and stuck to stereotypical routines to avoid getting lost in the network of display possibilities and complex menu structures.

Note how the concept of escalating demands as events cascade relates to the phenomenon of clumsy automation. Normally, the automation in these cases looked smoothly integrated, apparently reducing workload when situations fell within the textbook competence envelope represented by the model of operations, disturbances, and demands on operations implicit in the automation suite. But situations arise which fall outside this textbook envelope given the limits of that model (brittleness) and changes in the world. In these situations, the penalties for poor coordination (low observability, low directability) come to the fore. Ironically, practitioners' adaptations to make the system work hide the clumsiness from distant outsiders (Law of Fluency, p. 20).

TAILORING

Adaptation occurs in order for practitioners to cope with the complexities introduced by clumsy use of technology that is ill-fit to the demands of work (*JCS-Foundations*, pp. 106-107). In one type of adaptation—system tailoring, practitioners adapt the device and context of activity to preserve existing strategies used to carry out tasks (e.g., adaptation focuses on the set-up of the device, device configuration, how the device is situated in the larger context). In another type— task tailoring, practitioners adapt their strategies, especially cognitive and collaborative strategies, for carrying out tasks to accommodate constraints imposed by the new technology.

When practitioners tailor systems, they adapt the device itself to fit their strategies and the demands of the field of activity. For example, in operating room study practitioners set up the new device in a particular way to minimize their need to interact with the new technology during high criticality and high tempo periods (Cook & Woods, 1996). This occurs despite the fact that the practitioners' configurations neutralized many of the putative advantages of the new system (e.g., the flexibility to perform greater numbers and kinds of data manipulation). Note that system tailoring frequently results in only a small portion of the "in principle" device functionality actually being used operationally. Studies of JCSs in context often observe that operators throw away or alter functionality in order to achieve simplicity and ease of use.

Task tailoring is a type of adaptation where practitioners adjust their activities and strategies to accommodate constraints imposed by characteristics of the devices they use. For example, information systems which force operators to access related data serially through a narrow keyhole instead of in-parallel, result in new display and window management tasks (e.g., calling up and searching across displays for related data, decluttering displays as windows accumulate, etc.). Practitioners may tailor the device itself, for example, by trying to configure windows so that related data is available in parallel. However, they may still need to tailor their activities. For example, they may need to learn when to schedule the new decluttering task (e.g., by devising external reminders) to avoid being caught in a high criticality situation where they must reconfigure the display before they can "see" what is going on in the monitored process (Cook & Woods, 1996).

Task and system tailoring represent coping strategies for dealing with clumsy aspects of new technology. A variety of coping strategies employed by practitioners to tailor the system or their tasks has been observed. For example, when we observe that practitioners force device interaction to occur in low workload periods to minimize the need for interaction at high workload or high criticality periods, we are witnessing a general pattern in how people cope to prevent workload bottlenecks. When we observe that practitioners abandon cooperative strategies and switch to single-agent strategies when the demands for communication with the machine agent are high, practitioner adaptations have revealed automation that is poorly designed for the temporal flow and potential bottlenecks of work.

Another widespread observation is that, rather than exploit device flexibility, practitioners externally constrain devices via ad hoc standards. Individuals and groups develop and stick with stereotypical routes or methods to avoid getting lost in large networks of displays, complex menu structures, or complex sets of alternative methods. For example, Figure 13 shows about 50% of the menu space for a computerized patient monitoring information system used in operating rooms. Cook & Woods (1996) sampled physician interaction with the system for the first three months of its use during cardiac surgery. The highlighted sections of the menu space indicate the options that were actually used by physicians during this time period. This kind of data is typical—again, to cope with complexity, users throw away functionality in order to achieve simplicity of use tailored to their perceptions of their needs.

Studies of practitioner adaptation to clumsy technology consistently observe that users invent "escapes"—ways to abandon high-complexity modes of operation and to retreat to simpler modes of operation when workload gets too high. Finally, observations indicate that practitioners sometimes learn ways to "trick" automation, e.g., to silence nuisance alarms. Practitioners appear to do this in an attempt to exercise control over the technology (rather than let the technology control them) and to get the technology to function as a resource or tool for their ends (Roth et al., 1987).

Note that these forms of tailoring are as much a group as an individual dynamic. Understanding how practitioners adaptively respond to the introduction of new technology, and understanding the limits of their adaptations, are critical for understanding how new automation can create new vulnerabilities and paths toward system breakdown.

The patterns of practitioner adaptations reveal another important aspect of work in high consequence settings. Practitioners (commercial pilots, anesthesiologists, nuclear power operators, operators in space control centers, etc.) are responsible, not just for device operation but also for the larger performance goals of the overall system. Practitioners tailor their activities to insulate the larger system from device deficiencies and peculiarities of the technology. This occurs, in part, because practitioners inevitably are held accountable for failure to correctly operate equipment, diagnose faults, or respond to anomalies even if the device setup, operation, and performance are ill-suited to the demands of the environment (Cook et al., 2000).

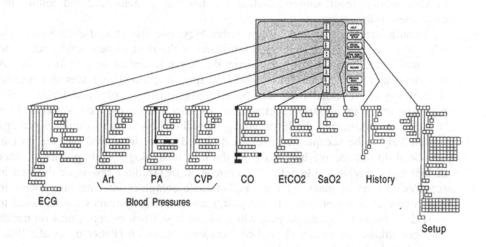

Figure 13 Menu space for an operating room patient monitoring system. The highlighted areas (black areas) are the items actually used by practitioners during observations of device use in cardiac surgery over three months. Note that the space of possibilities is very large compared with the portion practitioners actually used. (From Cook & Woods, 1996; see also Woods et al., 1994, Chapter 5).

However, there are limits to a practitioner's range of adaptability, and there are costs associated with practitioners' using coping strategies, especially in non-routine situations when a variety of complicating factors occur. Tailoring can be clever or it can be brittle. Adaptations to cope with one glitch can create vulnerabilities with respect to other work demands or situations. Therefore, to be successful in a global sense, adaptation at local levels must be guided by the provision of appropriate criteria, informational and material resources, and feedback. One function of people in more supervisory roles is to coordinate adaptation to recognize and avoid brittle tailoring and to propagate clever ones. This is an example of how JCSs always represent some balance between distant supervision, which can take a broad perspective but can miss critical factors that require adaptation locally, and local actors, who have privileged aspects to the actual situations, opportunities and constraints that challenge distant plans but can mis-adapt to handle these disruptions relative to larger goals and side effects.

The effects of adaptation in practice are not always directly evident. For example, at one point in time, practitioners adapt in effective ways based on the prevailing conditions. However, later, these conditions change in a way that makes the practitioners' previous tailoring ineffective, brittle, or maladaptive.

Paradoxically, practitioners' normally adaptive coping responses help to hide the corrosive effects of clumsy technology from designers and reviewers. Because practitioners are responsible, they work to tailor the technology to demands, making work appear smooth, hiding the underlying demands, and hiding the clumsiness from designers.

To anticipate the response to new technology, one also should examine closely who pays the costs in terms of new workload or the risks of new bottlenecks, and who receives the workload benefits from the new information or activities. Just as it is important to understand how workload spreads out and coalesces in time, the distribution of workload costs and benefits over different roles and levels of the organization is an important determinant of how people use new systems. Don Norman (1988), named the basic effect, "Grudin's Law:" New computer systems are unlikely to be accepted or used as designed when those who pay the cost in workload do not receive benefits from the new technology—only people in other distant roles receive the benefit. The operation of Grudin's law is easily seen in repeated failures of many attempts to introduce computer information systems in health care clinical contexts. In the past, many specific systems were designed to provide benefits to distant parties, while adding new work or constraints on health care practitioners who already risked workload saturation (Patterson et al., 2002; Koppel et al., 2005; Wears & Berg, 2005).

Adapting to the potential for workload and attentional bottlenecks should situations cascade and demands escalate is one of the constraints that drives adaptation in JCSs at work.

FAILURE OF MACHINE EXPLANATION
TO MAKE "INTELLIGENT" SYSTEMS TEAM PLAYERS

The concept of escalation helps us understand why efforts to add machine explanation to intelligent systems have failed to support cooperative interactions with human practitioners. Typically, such expert systems (or artificial intelligence—AI) developed their own solution to the problem at hand (user interaction focused on providing service to the algorithm so it could run the case at hand). Potential users found it difficult to accept such recommendations without some information about how the AI system arrived at its conclusions. This led many to develop ways to represent knowledge in such systems so that they could provide a description of how a system arrived at the diagnosis or solution (e.g., Chandrasekaran, Tanner & Josephson, 1989; Clancey, 1983).

However these explanations were generated, and however they were represented, these explanations were provided at the end of some problem-solving activity after the intelligent system had arrived at a potential solution. As a result, they were one-shot, retrospective explanations for activity that had already occurred. Developers who were busy creating or promoting autonomous machine analysis agents, and associated explanation-generating mechanisms and representations, didn't look for or notice the difficulties with explanations of this form. Developers didn't notice that the class of situations and contexts they assumed were standard were, in fact, quite over-simplified. Their model of practice missed many of the important constraints and demands, e.g., they assumed low tempo, low workload, low attentional demand situations, and they assumed that anomaly response consisted of mapping symptoms into a few diagnostic classifications—starkly inaccurate, given the actual character of anomaly response as we saw in Chapter 8.

But notice how people interact during anomaly response as in the update case (pp. 91-92). Exchanges are brief, coded, to the point. They do not break from the flow of the events and responses to engage in long-winded explanations (except when tempo drops low and then these exchanges take the form of tutorials from more experienced to less experienced practitioners). When people engage in collaborative problem solving, they tend to provide information about the basis for their assessments as the problem-solving process unfolds to build a common ground to support future coordination (e.g., Clark & Brennan, 1991; Cawsey, 1992; Johannesen, Cook & Woods, 1994).

Warnings about problems with one-shot, retrospective explanations were disregarded until AI diagnostic systems were applied to dynamic situations. As soon as such prototypes or systems had to deal with beyond-textbook situations, escalation occurred. The explanation then occurred at a time when the practitioner was likely to be engaged in multiple activities as a consequence of the cascade of effects of the initial event and the escalating cognitive demands for understanding and reacting as the situation evolved. These activities included generating and evaluating hypotheses, dealing with a new event or with the consequences of the fault(s), planning corrective actions or monitoring for the effects of interventions, attempting to differentiate the influences caused by faults and those caused by corrective actions, among others. This is captured in the mission control case described on pages 69-70 and summarized in Figure 7.

These kinds of expert systems did not act as cooperative agents. For example, the expert systems did not gauge the importance or length of their messages against the background context of competing cues for attention, or the state of the practitioner's ongoing activity (e.g., the machine agent could not assess another's current focus and judge when they were interruptible as in Dugdale et al., 2000). Thus, the system's output could occur as a disruption to other ongoing lines of reasoning and monitoring (Woods, 1995b). From practitioners point of view, the output of these systems was a false alarm often enough that the informativeness of their messages received dropped significantly (Woods, 1986a; Roth et al., 1987).

In addition, the presence of the intelligent system created new demands on human practitioners. The typical one-shot retrospective explanation was disconnected from other data and displays the practitioner was examining. This meant that the practitioner had to integrate the intelligent systems assessment with other available data as an extra task. This new task required that the practitioner shift attention away from what was currently going on in the process at just the time when it was particularly important to follow the cascade of events, thus possibly missing events.

Overall, the one-shot, retrospective style of explanation easily broke down under the demands of cascading situations. Practitioners, rather than being supported by the new systems, found an extra source of workload during high-tempo periods and a new source of data competing for their attention when they were already confronted with an avalanche of changing data. As a result, practitioners adapted. They simply ignored the intelligent system (e.g., Remington & Shafto, 1990, for one case; Malin et al., 1991 for the general pattern).

There are several ironies about this pattern of technology change and its surprising reverberations. First, it had happened before. The same experience had occurred in the early 1980s when nuclear power tried to automate fault diagnosis with non-AI techniques. The systems were unable to function autonomously (a case of Robin Murphy's law of developing for autonomy in action) and only exacerbated the data overload that operators confronted when a fault produced a cascade of disturbances (Woods, 1994; Woods et al., 2002). That attempt to automate diagnosis was abandoned, although the organizations involved, as well as the larger research community, failed to learn and share general patterns about coordination in anomaly response.

A second irony is that, to improve systems, expanding what one agent could do autonomously was insufficient. It proved necessary to consider the ways that cognitive activity is distributed and to design to support coordination (Hutchins, 1995a). The developers had assumed that their intelligent system could function essentially autonomously (at least on the important components of the task) and would be correct for almost all situations. In other words, they designed a system that would take over most of the work. The idea that human-intelligent system interaction required significant and meaningful cooperative activity that was adapted to the changing demands and tempo of situations, fell outside of the boundaries of their model of the demands of anomaly response and how actual fields of practice function.

WHY IS TECHNOLOGY SO OFTEN CLUMSY?

Running underneath the cases and patterns in this chapter is a question. The subtext is not simply patterns in JCSs, but moreover—Why are technological capabilities so often used clumsily in creating new complexities for already beleaguered practitioners? How do developers, and the blunt end context for development, mis-engineer JCSs so regularly? As in the last example of the failure of machine explanation, these cycles of mis-engineering JCSs at work have been repeated, revealing the same general patterns over again (Woods & Tinapple, 1999; Woods & Christoffersen, 2002).

Many of the techniques and patterns collected in this book represent partial answers in the form of correctives for discovering how JCSs work and for designing JCSs that work: balance multiple forms of observation, understand how JCSs adapt, use techniques for functional synthesis, respect general laws that govern JCSs at work, discover what makes work difficult, and more.

The processes of escalation also capture a pattern about the design of JCSs—it is a pattern in work that is difficult for design to see at work and to anticipate how to provide support in these situations. In canonical/textbook cases the technology seems to integrate smoothly into the work practices, bringing benefits. The practitioners are able to process information from machine agents. The additional workload of coordinating with a machine agent is easily managed. More static views of the work environment may be acceptable simplifications for situations falling within the textbook competence envelope. The penalties for poor design of supporting artifacts emerge only when unexpected situations dynamically escalate cognitive and coordination demands. In part, developers miss higher demand situations when design processes remain distant and disconnected from the actual demands of the field of practice. Hence the emphasis on field-oriented observation and the value of the struggle for authenticity in techniques to discover how JCSs work. In part, developers misread and rationalize away the evidence of trouble created by their designs in some scenarios. This can occur because situations that escalate are relatively less frequent than canonical cases. Also, because practitioners adapt to escape from potential workload bottlenecks as criticality and tempo increase, the necessary fluency of practice hides the evidence that the system does not fit operational demands (Woods et al., 1994).

Supporting the escalation in cognitive and coordination activity as problems cascade is a critical design requirement for innovating or developing JCSs that work. This illustrates how finding patterns in how JCSs work results in defining generic requirements to be supported through design. These design requirements are not about artifacts as an object, rather they are about how such objects change the function of JCSs (Woods, 1998).

To cope with escalation as a fundamental characteristic of cognitive work, one needs to consider how more knowledge and expertise need to be brought to bear on a cascading situation, and how more resources can be synchronized to handle the escalation of monitoring, attentional, and other demands (like the updating or coming up to speed processes in the case on pp. 91-92 and in Patterson & Woods, 2001). Functional models such as goal-means decompositions (Rasmussen et al., 1994) are valuable here because they help capture how situations cascade and how dilemmas such as goal conflicts arise.

Because escalation is fundamental to the work of JCSs, it specifies one target for scenario or problem design (Carroll, 1997). Natural history techniques, such as building and analyzing corpuses of critical incidents, are needed to understand how situations move from textbook to non-routine to exceptional in particular fields of practice, and particularly how these processes change following significant organizational or technological changes. Work is needed to identify general and specific complicating factors that shift situations beyond the envelope of textbook plans (Woods, 2006a).

The concept of escalation is not simply about a world of work, or about people's strategies, or about technological artifacts. Rather, it captures one kind of dynamic interplay between all these factors as it illustrates demands for coordination, resilience and affordance. As a result, escalation captures the fundamental point in Cognitive Systems Engineering—the joint cognitive systems is the fundamental unit of analysis for progress on understanding and designing systems of people and technology at work (*JCS-Foundations*, Chapter 6).

MAKING AUTOMATION A TEAM PLAYER

Consider this example from a study of pilot interaction with cockpit automation (Sarter & Woods, 2000).

What did the automation do?
A pilot prepares his descent into the destination airport and receives an initial ATC clearance for an instrument landing approach to runway 24 L together with a number of altitude constraints for various waypoints of the arrival. The pilot programs these constraints into the flight automation.

Shortly after the entire clearance has been programmed, an amended clearance is issued by ATC to now make an instrument landing approach to runway 24 R. (The change in runway was made because of an airport equipment failure.) When the pilot changes the runway in the instructions to the automation, the automation signals that it understands the runway change and begins to act based on this new target.

Question: Does the automation continue to use the altitude constraints *which still need to be respected* in this situation?

Pilots tend to assume that the automation will remember the previously entered altitude constraints. However, the automation is designed to delete the previous entries when a new runway is entered. Unless the pilot notices that the indicators of the altitude constraints are missing from displays and re-enters the constraints, the automation will fly the aircraft through these altitudes. See Sarter & Woods (2000) for the results on how line pilots handled this situation in a Scaled World study.

If the pilot were working with a human team member, it would be reasonable to expect that person would continue to apply the constraints—to "remember" the previous intent and know that it is still relevant to the situation at hand. Both human team members know that the altitude constraints apply to the current situation, and both know that the change in runways is not relevant to the altitude constraints. However, in this instance the automation doesn't behave this way—the software drops all of the altitude constraints entered for the originally planned approach. In addition, the only signal that these constraints have been dropped is through the *disappearance* of two indicators that would be present if the constraints were still active and governing how the automation will fly the aircraft.

The pilots may not understand or anticipate that the automation does not continue to respect the constraints following a change but rather reverts to a baseline condition.[11] This lack of inter-predictability (and the lack of observability in the displays of automation activity) creates the conditions for an automation surprise—the automation will violate the altitude constraints as it flies the descent

---

[11] Designers' ability to decide on a general policy for how the automation should behave given this change in plan is limited. There could be situations where the preference would reverse (where the constraints are no longer relevant following the change). The issue raised by the example is that it is not obvious to the human partner how the automation handles the change or how to modify its behavior if its default is inappropriate for the situation at hand. Plus, even if one understands how the automation would behave in this situation, would this knowledge help anticipate how the automation will behave in other apparently similar or related situations from the point of view of human practitioners?

unless the human pilot notices and intervenes. In the Woods & Sarter (2000) scaled world study, 4 of 18 experienced pilots never understood or noticed the automation's behavior and 12 of the 14 who noticed at some point were unable to recover before the constraints were violated.

This is a vulnerable point for coordination breakdown in one system and work domain, but it also illustrates several points about human-automation teamwork in general[12] (Woods & Sarter, 2002; Klein et al., 2004). Increases in machine autonomy do not remove the designer's responsibility to provide support for basic functions in coordinated activity. The first step in meeting responsibility to design for coordination is to make automation activities *observable*. In the above case, it is not obvious what the automation is going to do following the runway change to the approach plan—relying on people to notice the absence of indicators is not a good way to achieve observability. The second step is to provide *directability*—the ability to direct/re-direct resources, activities, priorities as situations change and escalate. In this case, as in the 1999 Global Hawk accident (pp. 113-115), the ability to re-direct the automation is limited, given the high tempo of the situation as evidenced by the fact that few pilots were able to reprogram the automation before the altitude constraints were violated.

As these situations may occur infrequently, how will the pilot know when to check up on the automation's future behavior given the pilot's need to accomplish other tasks and monitor other aspects of the situation? Without support for the function of *directed attention*—the ability to re-orient focus in a changing world—the pilot's attention reasonably could be focused elsewhere, without the awareness that there is a need to check what the automation will do next.

However, the case illustrates a deeper concern about people and automata. The example illustrates how there is a basic asymmetry between people and automata in terms of the competencies that support coordination (Klein, Feltovich et al., 2005). Coordination depends on the ability to anticipate the actions of other parties, and to participate in coordinated activity means that each party has a responsibility to make his or her actions sufficiently predictable to others—*inter-predictability*

This discussion raises the question of whether automata can be predictable. How does experience with the automation in other situations help a pilot anticipate its behavior in the case of the late runway change? How does experience with the automation's behavior in this case help one generalize to other situations? There is nothing in the algorithms themselves that helps another to generalize how the algorithm will behave in other situations—one must either know the details of the automation's algorithms or have direct experience with the system in this detailed situation to know how the automation will behave.

Woods (1996) saw this as the paradox of perceived animacy of automata. Autonomy and authority are properties that convey an agent-like status on automata

---

[12] Also note how the case illustrates general principles in the design of problems/scenarios and scaled world studies. The probe event—late runway change—is not simply a specific glitch to be repaired in incremental design; its serves as an instance of class of events that challenge coordination and test how well the JCS meets requirements to support observability, directability, inter-predictability, and directed attention in general.

from the point of view of outside human observers. The perception of agency (and animacy) concerns the relationship between the behavior of a system: what can be seen about that behavior by human observers relative to other observable contingent events (Heider & Simmel, 1944; Bradshaw, 1997; Scholl & Tremoulet, 2000). As a result, automated systems, as they increase in autonomy (varying degrees of capability to carry out sequences of actions independent from human input), authority (the power to take-over control of a process), and complexity (more extensive and tighter coupling), have two kinds of interpretations: as a deterministic machine in hindsight and as an animate agent in context capable of activities independent of the operator.

Automated systems are completely predictable in principle when based on deterministic algorithms. If one has knowledge of outcome, no time pressure, complete knowledge of how the system works, complete recall of the past instructions given to the system, total awareness of environmental conditions, then one can project accurately the behavior of their automated partner or can retrospectively show how a system's behavior was deterministic.

However, those who monitor or interact with automata in context may perceive the system very differently because, as the system becomes more autonomous and complex, projecting its behavior also becomes more challenging. A user's perception of the device depends on an interaction between its capabilities and the feedback that reveals or makes *observable* the workings of a process or device in relation to events in the environment. What feedback is available depends upon the "image" the device presents to users (Norman, 1988). When a device is complex, has high autonomy and authority, and provides weak feedback about its activities (low observability), it can create the image of an animate agent capable of independent perception and willful action. (For a different approach to how people interpret the behavior of automata, see Reeves & Nass, 1996; Nass & Moon, 2000).

As a result, an automated system can look very different from the perspective of a practitioner in context as compared to an analyst taking a bird's eye view with knowledge of outcome. The latter will see how the system's behavior was a direct and natural result of previous instructions and current state; the former will see a system that appears to do surprising things on its own. This is the paradox of the perceived animacy of automated systems that have high autonomy and authority but low observability (Lanir, 1995, p. 68, put it more entertainingly, defining "*agent*: n. A computer program whose user interface is so obscure that the user must think of it as a quirky, but powerful, person...").

The animacy paradox occurs because how one sees and interprets a machine's behavior depends on one's vantage point—in hindsight and with complete knowledge of circumstances and of the device, the system looks determinate ("the system performed as designed"; and see the list on pp. 123-124); in foresight, the system is considered animate because it seems to act willfully and surprisingly on its own ("What is it doing now? Why? What will it do next?" and the list on p. 119).

The above discussion raises deep questions about limits on how automata can participate in coordinative activity. These limits indicate that designers and the automata's human partners are responsible for making up for these deficiencies

(Klein et al., 2004). Designers need to provide means to support coordination by building in the means to help people to anticipate the automata's behavior and redirect the automata's activities to meet or balance the relevant goals.

To expand our view of the demands in coordination, let us consider vulnerabilities and expertise in coordinated activity across people in different roles as they attempt to replan to handle a disrupting event.

A Coordination Breakdown in Response to a Disrupting Event

When coordination is smooth, the effort involved is almost invisible (an example of the Law of Fluency). Occasionally, breakdowns in synchronization and communication occur between the players. These breakdowns provide the opportunity to see the mechanisms for coordination at work in the JCS. Consider a case of a communication breakdown during the modification of a plan to see some the factors involved in distributed replanning. The case is drawn from the air traffic management system circa 1998, which is one example of a distributed system in which plans in progress are modified to meet changing circumstances. In air traffic management, airline dispatchers and flight crews work with air traffic controllers to adapt flight plans to accommodate disruptions in context; for example, to handle weather created obstacles in the planned flight path.

Replanning Breakdown:
The airline dispatcher in charge of a flight filed an initial flight plan from Dallas/Ft. Worth to Miami. After the flight was on its way, the dispatcher noticed a line of thunderstorms moving into the aircraft's planned route, and, with the Captain's concurrence, issued a reroute instruction to the aircraft. During this process, the Captain was informed about the situation and the weather conditions that prompted the reroute. The rerouting was coordinated with Air Traffic Control (ATC) and was approved. The dispatcher in charge of the flight, who originally noticed the bad weather, was also responsible for about 30 other planes. Once he had filed the reroute, briefed the Captain, and obtained permission from the first ATC center, he felt that the issue had been resolved, and devoted his attention to other airplanes and tasks in his scope of responsibility.

However, as the flight progressed, the Receiving Center, (i.e., the ATC center that was taking on responsibility for the flight as the aircraft transitioned from one ATC sector to another) could not accommodate the reroute due to traffic along the east coast of Florida, and put the airplane back on its original flight plan. The Captain assented, assuming that the receiving ATC center also knew about the weather front. As the weather system developed, following the original flight plan trapped the aircraft south of the line of thunderstorms, forcing the aircraft to circle, waiting for a break (airborne holding). As the aircraft began to run low on fuel, the crew faced an undesirable situation with limited options in terms of diversion airports (the weather either affected its alternative airports or blocked their path to them). As a consequence, the crew faced a situation of being very low on fuel, with limited options in terms of available diversion airports. The aircraft finally broke through the line of thunderstorms as the weather passed south of Miami, and was able to land there, though the aircraft passed through severe turbulence (Smith et al., 2001).

Note how the disrupting event was recognized and replanning begun (Smith et al., 2003). Each part of the distributed system had partial knowledge relevant to the situation and constraints. Each attempted to carry out their role effectively relative to the disruption represented by the line of thunderstorms. All of the responsible parties communicated, building the sense that all were operating on a common assessment of the situation, that all had examined the issue at hand, and that all were working to the same subset of goals. Yet each of the parties misunderstood the others' perspectives, failed to correct miscommunications, failed to integrate information and coordinate across the different fields of view and scopes of responsibility. For example, the second ATC had a different set of tasks and a different focus of attention (north–south routes on the east coast of Florida), and did not have complete data about the weather situation in the region that concerned the pilot. Neither party realized that they were on different wavelengths and neither made any inquiries to verify that their assumptions about the other's knowledge, goals, and activities matched their own. Meanwhile, the third party to the plan modification, the dispatcher, had reached closure on this issue and did not realize that it had been reopened in time to help. This left the flight crew to deal with the obstacle of the weather system—an obstacle that the replanning process had anticipated and tried to avoid. In a sense, the miscoordination managed the flight into a much more difficult situation (Woods & Sarter, 2000).

This case captures portions of the process of modifying a plan in progress portion of anomaly response shown in Figure 11 (Potter et al., 1996). The original plan occurs remotely in time and space from the execution of it. In this gap, surprises, i.e., events that challenge the assumptions and goals of the original plan, such as the changing weather system, can develop. This disrupting event is recognized, and initiates a round of analysis and consideration of modifications (Chow et al., 2000). This involves communication across the multiple parties involved in the plan (ATC, flight crew, and company dispatch) each with different data, knowledge, assessments, goals, and responsibilities. There are points where handover of responsibility occurs (the aircraft transitions between ATC sectors). These points are places where reconsideration of the modified plan can occur.

To smoothly resolve the disrupting event through replanning, a variety of forms of distributed work needed to occur (Smith et al., 2001; 2004):

• Each group needed to step outside its usual role to see the situation from other groups' perspectives.
• Each group needed to integrate information and knowledge across multiple groups to create a complete picture larger than what any one group saw.
• Each group needed to recognize how the multiple relevant constraints were combining to create a squeeze on this flight.
• Devising a new plan required accommodating the multiple constraints operating in this situation that cut across the different groups and roles.

This story of coordination arises from a demand for resilience, and the miscoordination observed illustrates the brittleness of the JCS in that work domain at that point in time[13] (see Woods & Cook for other incidents that reveal sources of resilience and brittleness).

Stories of coordination and miscoordination such as those in this chapter reveal some basic laws that need to be respected when designing for collaborative work. First, collaborative work occurs when the agents coordinate activity *in the process of* recognizing and solving a problem. Or to state the corollary: "It's not cooperation if either you do it all or I do it all" (Woods, 1986a; Roth et al., 1987). Second, collaborating agents have access to partial, overlapping information and knowledge relevant to the problem at hand. No one agent or group in the situation has an all-encompassing perspective that covers all of the ongoing events at the correct scale, possesses all of the information relevant to the evolving situation, or has access to all of the knowledge and constraints that need to be considered in revising the plan given the disrupting event. The power of coordinating perspectives in work is the expansion in field of view, greater access to relevant information, and broadened knowledge of the constraints to be considered. However, the case illustrates a third general principle about collaboration. Technologists often mistake connectivity (the technical capability to connect to disparate parties and data sources) for coordination. In other words, it is easy to assume that coordination comes for free if only sufficient connectivity can be established. However, observation quickly reveals that achieving coordinated activity across agents consumes resources and requires ongoing investments and practice—*coordination costs, continually* (Klein et al., 2005; see Patterson et. al., 2002, for a case in health care or Olson et al., 2001).

---

[13] Air traffic control in the U.S. has made major strides in applying lessons of research on coordination in distributed systems to improve flexibility and efficiency as the air traffic management system was re-designed by the FAA's Air Traffic Control Systems Command Center and Collaborative Decision Making Program (see Smith et al., 2004).

To summarize: Automation Surprises

To study and design work, the unit of analysis is the joint cognitive system—not people and computers as separate components which then interact. Observing JCSs at work reveals processes of coordination and miscoordination as multiple parties contribute to handling evolving situations and achieving goals. Automation surprises are not simply a type of coordination breakdown at the sharp end of practice. The stories of coordination breakdown reveal a deeper surprise about automation as developers' oversimplifications about how JCS work lead them to miss predictable effects of technology change that create new complexities (workload, attention bottlenecks), miss how people adapt to work around complexities, and miss the actual functions in coordinated activity that require support.

Additional resources: Aviation conducted a natural experiment by introducing new levels and forms of cockpit automation and then investigating to understand the surprising reverberations. Many sources are available on this process of technology change, surprising effects, and industry-wide adaptation to reach a new equilibrium such as Billings (1997) and the FAA Human Factors Team report (Abbott, 1996). Failures of automated systems in health care also provide a rich set of resources to buttress the ideas and studies presented here, and in health care there is no better example of clumsy automation than infusion devices (Cook et al., 1998). Success stories can be found in work on side effects of introducing bar coding medication administration systems (Patterson et al., 2002), critiquing support for antibody identification in blood banking (Guerlain et al., 1999), coordinating distributed replanning between airlines and FAA in air traffic management (Smith et al; 2001; 2004). Stories of clumsy automation are balanced by a dramatic increase in studies that show how collaboration works well, often in various control centers. Understanding the basis for coordinated work in space shuttle mission control is one resource (Patterson et al., 1999), but many are available (e.g., Heath & Luff, 1992; Olson et al. 2001).

Chapter 11

On People and Computers in JCSs at Work

This chapter shifts perspective to consider beliefs about JCSs at work. The chapter uses a recent shift in new technology—robotic platforms that can move on their own in the world—to show how CSE participates in the processes of envisioning future operational worlds given opportunities defined by new technology. In this story about envisioning, concepts about how JCSs work are revealed as well— concepts about: how contrasts across multiple perspectives aid in re-conceptualization, how people and only people can be responsible agents—problem-holders—balancing conflicts across multiple goals, and how automata as literal-minded agents require people to close the context gap.

In CSE, the concept of the Joint Cognitive System becomes the critical unit of analysis for work systems. The opposition, separation, and substitution of people and machines disappears (Woods & Tinapple, 1999). The goal is to design for coordination and resilience of the joint system as it adapts to the demands of work and adapts artifacts to support strategies for work. This is in stark contrast to the repeated attempts to design and embody autonomous algorithms as means to overcome human limits and to substitute these for human involvement (*JCS-Foundations*, p. 17). Historically, the substitution approach has been encapsulated through (cf., *JCS-Foundations*, pp. 121-123):

• Continuing attempts to update lists of what machines do well versus what people do well (based on the original attempt by Fitts, 1951, hence the name Fitts List; cf. also, Hoffman et al., 2002);
• Continuing attempts to define different degrees of substitution between people and automation given different levels of autonomy and authority of machines (e.g., Sheridan 1992; *JCS-Foundations*, Figure 6.2, p. 119);
• Continuing attempts to design systems based on allocation of tasks between people and machines, which assumes decomposability of work into independent parts or tasks.

Each round of technology change based on these assumptions has produced quite surprising effects and reverberations (e.g., Table 5). When confronted by evidence that these assumptions are wrong, the response usually takes the form of continued pursuit of the Holy Grail of isolated autonomy with the promise that the next time the technology advances will provide the predicted effects and only those effects—"Just a little more autonomy will be enough the next time."

All of these tactics represent over-simplifications in the face of the complexities of joint systems (as captured in Table 4). These tactics attempt to decompose a dynamic, adaptive, multifactor process—the JCS—as if it could be treated as a collection of independent parts, handling situations decomposed into a series of discrete snapshots. Studying and designing work systems based on these over-simplifications results in another kind of automation surprise for those at the blunt end of systems—as they mis-assess the effects of each episode of technology change (Woods & Dekker, 2000).

CSE as a cognitive *Systems Engineering* uses the system perspective heuristics described on p. 7 to escape from the risk of the above oversimplifications:

(a) shift to a broader perspective by drawing a wider boundary to define the system of interest—the focus on the joint cognitive system;
(b) synthesize the relationships that emerge when modeling based on this broader perspective—observe and model phenomenon of resilience (adaptation), coordination, and affordance;
(c) adopt a "middle-out" approach to take into account cross-scale interactions— how coordinated activity to meet multiple demands and goals (the "sharp end" of practice) sets the context for a more "micro" level of analysis of private cognition and is constrained by the larger context of organizational dynamics at the "blunt end" (pp. 8-9, Figures 2 and 3).

Let us examine how three different perspectives anticipate the effects of changes made possible by a specific wave of new technology: the impact of introducing mobile robotic systems into demanding operational contexts such as search and rescue, military operations in urban settings, space exploration, and coordinating multiple unmanned aerial vehicles and unmanned ground vehicles—UAVs/UGVs (Bradshaw et al., 2004; Murphy, 2004).

ENVISIONING THE IMPACT OF NEW TECHNOLOGY: STORIES OF FUTURE OPERATIONS WITH ROBOTS

As robotic capabilities expand, many stakeholders attempt to anticipate the organizational and technological changes these capabilities will produce. This is a case of envisioning the future of operations if new artifacts are realized and deployed (Woods & Dekker, 2000). These kinds of design envisioning tasks are difficult as different specialists each have only partial views of the space of design possibilities and the potential future impact of alternative design directions. In this case we explore what happens when three different perspectives—a roboticist, a cognitive systems engineer, and a problem-holder—confront each other while envisioning the impact of robotic systems on the future of demanding work settings (Woods, 1998; Woods & Christoffersen, 2002; Woods et al., 2004).

The first character in this episode is the *problem-holder*, who also acts as a *reflective practitioner* in the discussions (i.e., he has extensive experience in the field of practice—practitioner, but also has some ability to step outside of that role and analyze the nature of his and his colleagues work—reflective; see Schön, 1983). His perspective focuses on how to meet pressing needs though timely introduction of new systems into the field of practice. Being responsible for meeting the pressing need makes this character the problem-holder. Demands for new levels of performance and pressures to be more efficient with resources lead organizations to make significant investments in new rounds of development (and, if necessary, research) to field the infrastructure for new operational systems (e.g., soldier-robot teams). As a leader in this development process, this character is under a perceptible and omnipresent shadow to demonstrate progress toward fielding new systems in a reasonable time frame. As a reflective practitioner, he represents a development process grounded on direct experience with the real difficulties of the operational setting, such as planning for urban military operations or past search and rescue deployments.

The second character is a *roboticist,* who represents the goal of advancing what machines can do autonomously (advancing the technology baseline). In advancing the autonomous capabilities of robotic systems, human interaction issues arise only in terms of (a) building interfaces for remote humans to communicate, guide, instruct or takeover control with robots, and (b) social/organizational consequences of advances in autonomous agents. The former facet of this perspective puts advances in autonomous systems first and then considers questions about interfacing to people as a grudging, residual and temporary need for fielding the new capabilities in the short term. The latter facet of this perspective anticipates the technology trend toward increasingly autonomous machine agents and begins to contemplate what it means to introduce capable artificial but person-like agents into the human sphere (e.g., Reeves & Nass, 1996; Lopes et al., 2001).

For the technologist, the demands and pressures placed on fields of practice become windows of opportunity for investment in expanding the autonomous capabilities and in deployment of these capabilities. By promoting the potentially available increases in autonomous power, the technologist is shaping stakeholders' expectations and therefore demands on future operational performance. In envisioning the future, this perspective takes for granted that it is self-evident to all that the autonomous capabilities under consideration have significant utility. The mindset becomes centered on how to create the power and how to realize fieldable systems based on those powers.

Our third character is the *cognitive systems engineer* who represents the efforts to anticipate how to-be-realized systems may introduce new capabilities and new complexities and to anticipate how the JCS will adapt to exploit or workaround those effects. To assist in envisioning future cycles of adaptation, he draws on patterns about coordination, resilience and affordance abstracted from observations in many different situations. The goal of envisioning adaptation is to identify promising directions to support the changing demands of work—developing hypotheses about what would prove useful.

The different perspectives view prototypes differently (Woods, 1998; Woods & Christoffersen, 2002). For the roboticist, prototypes represent *future technological powers* (what could be available in the technology "store" for stakeholders to deploy). For the problem-holder, prototypes represent *partially refined final products*—the system that will be fielded as processes of detailing and glitch identification/repair are completed within schedule pressures. For the cognitive systems engineer, prototypes represent *tools for discovery*—promising hypotheses in the search to discover what will be useful (Perkins, 1992). The evidence or feedback that guides search in the discovery process is found in the adaptive response of the field of practice to changes represented or enabled by the introduction of new artifacts—how practitioners and others will adapt to exploit new capabilities or workaround new complexities.

The Scene

The scene opens with briefings to ground the participants on the functions and difficulties in the setting of interest. The fields of practice considered here—urban operations, search and rescue, and chemical/biological incident response—represent changing resource pressures and changing demands on performance. For example, the military are under new pressures where adversaries are embedded in urban infrastructure and populations. In the case of emergency first responders, new pressures arise for search and rescue operations with the possibility of injuries from chemical, radiological, and biological incidents (e.g., consider the Tokyo sarin gas attack in 1996 and the Russian rescue of hostages in a Moscow theatre in 2002).

Grounding the interaction across the three perspectives on the performance demands and characteristics of these settings enables a discussion of team play and coordination in evolving situations. As the discussion builds around the briefings, each character looks for an opportunity to make an opening statement that captures their stance and role in the envisioning process. Either literally or implicitly, the roboticist begins with—"I want to talk about autonomy...," the problem-holder responds—"I desperately need help to meet real and pressing demands...," and the cognitive systems engineer interjects—"I want to talk about adapting to changing complexity... ."

The statement—"I want to talk about autonomy..."—reveals that the mindset of the roboticist is *how do the needs of the problem-holder/work setting connect to the advancing capability for what robots can do autonomously*? The discussions are viewed through the filter of questions such as—how do I get robots to do that? We can get robots to do this, would that help? He tries to understand the desires/needs of the problem-holder and reformulate those statements in terms of the maturing capabilities and constraints of robotic technology in mobility, sensors, and communication. The discussion points he raises center on the pace and character of the advancing technology baseline.

Grounding the discussion on a scenario eventually raises questions about interactions across people and robots. These interaction requirements are viewed as another set of drivers on autonomous capabilities of the robot (can I get the robotic systems to exchange communications and updates to/from remote people). His interest focuses on how to resolve tradeoffs and constraints in robot capabilities versus task demands (tradeoffs created by power limits, bandwidth limits, size, range, etc.). Envisioning future robots for such teams requires balancing these interacting constraints; therefore, he focuses the conversation on trying to better understand the performance demands on the robotic platform. Practically, the roboticist is figuring out what kinds of working systems could be supplied within varying time/resource constraints. He considers how to integrate capabilities *and limits* across sub-areas of sensing, mobility, and communication (e.g., range limits on wireless communication) into a system that could be fielded into the operational setting (or, other words, *what technical advances are needed*).

The backdrop for these discussions is the assumption of future ubiquity and impact of robotic systems, e.g., in x years, robots will be generally accessible, effective; therefore, of common experience, and dramatic in impact on human roles. Given these assumptions, discussions of the relationship between people and robots can lead to broader discussions on the relationship of people to robots as one or another kind of nonhuman "persons." Robots, as agents that can move (and more) on their own in the physical world, raise questions about what is an agent, what makes for animacy, and what are the implications of introducing artificial persons into the human sphere.

The response—"I desperately need help to meet real and pressing demands…"—reveals that the mindset of the problem-holder centers on the new pressures where adversaries are embedded in infrastructure and populations (asymmetric urban operations), and highlights new capabilities such as technical rescues in which the team needs to stabilize a damaged structure and stabilize the injured person's physiology as part of the process of extraction from a collapsed building. The problem-holder considers particular difficulties that need to be overcome: "I want to talk about how to get across the street in urban combat, rescue a wounded comrade, enter a room with possibly hostile people but also with civilians mixed in." Or, "I want to talk about how to determine quickly the appropriate care needed by those injured when chemical/biological agents may be involved, extend the time human personnel can conduct search and rescue operations in a chemical/biological hot zone, and recognize and initiate care to the injured as they are transferred from the hot zone to decontamination stations."

She tries to steer the envisioning process by referring to or playing out stories that illustrate the difficult demands that arise and the strategies that have evolved to meet these demands. These stories emphasize how people work in teams and as units, not merely as individuals, to coordinate activities, adapt to surprises, and achieve goals.

As the roboticist lays out advancing capabilities, the reflective practitioner tries to consider how to translate these items (e.g., new sensors, the mobility envelope of micro-UAVs) into the operational structure of their organization and the roles/skills of the people who make up the team. As the reflective practitioner/problem-holder considers the match of robotic capabilities with demands, she realizes that there are new forms of workload as people need to coordinate with these kinds of robotic systems. With limited slack available in the current team organization, she asks, "Who is going to work with this robot?" "The personnel are all busy already so there is no one to run this?" The notion that the robotic system will simply replace some person or will substitute for a person in some function seems to miss how the current staff do the work together, intertwine and shift roles fluently, coordinate activities, and rely on each other in difficult and demanding situations. As a reflective practitioner, she tries to get across the mindset of good rescue teams—team members rely on each other, trusting each other with their lives. How does a robotic system fit into this atmosphere? Is it a reliable partner? Can I re-direct its activities? Is its behavior predictable across the range of situations that they might face?

As these two characters interact, connecting the two mindsets is quite difficult. The problem-holder tends to focus on incremental improvements to the current search and rescue activities; to the roboticist these seem to miss the revolutionary potential of robots as they become capable of more sophisticated forms of activity on their own. But when the roboticist points to the prospects for future autonomous capabilities, the problem-holder/reflective practitioner has great difficulty connecting these capabilities into the detailed practices and constraints on their work in actual situations.

It is at points like this that the cognitive systems engineer tries to shift the discussion, asking in various ways—"Let's talk about adapting to complexity... ." The cognitive systems engineer tries to probe the practitioner's experience base and, since this a reflective practitioner, to explore her models of what makes situations difficult in urban or rescue operations. What kinds of surprises and adaptation are needed? How do current teams achieve resilience and robustness? How will practitioners compensate for brittleness and other limits to automata?

He uses patterns from the research base on coordination, resilience and affordance as possible storylines. He considers how different examples of general patterns about JCSs may be at work in this setting, questions such as: How do events cascade and demands escalate as situations evolve? How do bottlenecks in workload, data overload, or demands on attention arise? How can one build a coherent assessment (and revise it) from partial data and views coming in over time? How do people build common ground in order to avoid coordination surprises? How do teams adapt when plans are disrupted? Can the system gracefully degrade or reconfigure quickly when assets are lost?

For example, the cognitive systems engineer notes that behind the discussion of first response scenarios a topic arises frequently for both rescue personnel and for robots—limits on energy/time consumption in access and egress to the scene of contamination influence many aspects of emergency response. Energy constraints on time in the hot zone become a candidate in the cognitive systems engineer's search for what might be dominating constraints on practice. The time/energy constraints help identify tradeoffs that make search and rescue difficult: getting in quickly to assess and do triage, given there are risks of delay to the victims and risks from speed to the rescuers; the need to quickly move the injured out of hot zones to decontamination stations and treatment, but prerequisite prior to moving injured people to gather more information about their injuries and about the sources of contamination present. Noticing these constraints and tradeoffs re-focuses the search for how robotic platforms and other technology could provide affordances to those responsible for emergency response (note that both the problem-holder and CSE perspective do not stop at one technology or one way of bringing a new technology to operations).

By identifying demands factors, the cognitive systems engineer can begin to project how the changing capabilities will transform the nature of practice: How will the changes introduce complexities to be worked around? Which capabilities will be exploited by leaders to accommodate new performance demands and resource pressures? What are the side effects of change that will need to be accommodated. The cognitive system engineer listens very intently to the discussions of the up coming technology changes because this information helps him anticipate the kinds of coordination that will characterize the future operational world and how human roles will change with new automata (e.g., how new forms of automata could create workload peaks at high tempo or high criticality periods).[14]

Talking in Synchrony?

Initially, there is a natural tendency for each of these perspectives to spin inwards focusing on their own role and activities in isolation.

• The cognitive systems engineer is sketching scenarios for exploring coordination in human-robot teams, and debating how prototypes could be used as tools for discovery of what would be useful (how CSE begins with gathering information to shape where and how to observe).

• The problem-holder focuses on learning as much as possible about the growing but concrete capabilities of robots to determine how these might help her escape the traps and dilemmas embedded in the operations and missions she will be asked to carry out (what is reliable and fieldable).

• The roboticist is pondering how to reconcile the varying competing constraints to match robot autonomous performance levels with scarce resources and how to integrate component capabilities/limits into a robotic system adapted to the demanding performance targets and limited budget required by this application.

---

[14] Fundamentally, CSE methods and concepts are about the future of work. Any methods or results in CSE are tested against the challenges to inform design (p. 58). CSE arose to move beyond methods that first required partial specification of tasks and artifacts in order to describe the activities required to carry out the emerging tasks and complete the specification of the artifacts which are to be used in those tasks. Development of large systems always requires such analyses as part of the specification process to detail fieldable devices, but the role of CSE is to support finding would will be useful (Woods, 1998; Roesler et al., 2005). Analyses for specification are not a form of cognitive task analysis and have nothing to do with the origins, needs, results, or promise of CSE. As CSE methods have come into broader use, some claim functional modeling only is possible for existing forms of work. This is a stunning misunderstanding of the whole point of functional modeling and synthesis which specifies a program for innovation for only partially decomposable systems like JCSs, as discussed in Chapter 6. The story of collaborative envisioning in this chapter is meant to illustrate broadly how the CSE perspective is essential to avoid getting lost as processes of change and adaptation accelerate.

The question is *how to cross-connect the different perspectives to avoid the fragility of envisioning future operations and to synthesize new promising insights.* When the different perspectives begin to synchronize, points of contact emerge. For example, from Woods et al. (2004):

Problem-holder: "What obstacles can it clear?"
Roboticist: "It can go over items 15 inches or less."
Cognitive systems engineer: "How do (would) you tell?"
Practitioner: "We drive up to the obstacle and if it's higher than the treads we know it cannot be scaled."
Cognitive systems engineer: "The practitioner's heuristic is an example of workarounds and inferences people develop to make up for the impoverished perceptual view through the robotic platform's sensors. In contrast, when people are physically present in the environment being explored, they pick up these affordances immediately."

This interplay triggers consideration of a broad set of factors that affect movement over broken terrain, situational variables that complicate clearing obstacles, and ambiguities in how remote observers perceive environments through a robot's sensors. New challenges that cross normal disciplinary boundaries and emphasize new connections are noticed. These challenges throw new light on questions of fusion across sensors, enhanced visualization concepts for remote human observers, reasoning about risk taking, and more.

Overall, broad findings emerge when the three perspectives begin to synchronize, such as:

• Many types of tradeoffs must be respected and balanced.
• Automata take direction and inform distant parties of local conditions.
• From the practitioner's point of view, the robotic system is a resource with limited autonomy.
• Being a team member includes the ability to pick up and adapt to the activities of others in the team to achieve coordinated activity.
• The target field of practice is extremely demanding, and stresses the resilience of any unit (set of agents) organized to perform as an adaptive team.
• Inevitably, autonomous capabilities will exhibit brittleness as situations develop beyond their boundary conditions.
• The difficulties of balancing the multiple constraints and tradeoffs highlights the adaptability of people, given sufficient training and practice.
• Human capabilities that support high levels of coordination can be used as a competence model to stimulate new ways of using the wide and expanding technological possibilities.

Synchronizing the perspectives requires the ability to translate between the languages of the practitioner (the risk of getting lost in the details of substantive fields of practice) and the languages of the technologist (the risk of chasing the tail of changing technological powers). Concepts about resilience, coordination and affordance in JCSs provide these needed cross-connections (Figure 1).

This story about envisioning the future of work also captures another aspect of JCSs. Each character in the story has access to partial information and knowledge relevant to the problem at hand. Success in design envisioning, as in all JCSs at work, depends on the ability to integrate and synthesize across these contrasting perspectives.

RESPONSIBILITY IN JOINT COGNITIVE SYTEMS AT WORK

Problem-Holders

How does responsibility for the consequences of actions influence the design of JCSs at work? Billings (1997) has developed a set of first principles for responsibility in human-automation systems, which builds from a basic premise (p. 39): *Some human practitioners bear ultimate responsibility for operational goals.* As a result, those with responsibility within the system must have some means to effectively command within that scope of responsibility (as problem-holders): *These supervisory human operators must be in command.* The question, then, is what does it mean to be "in command" of machine agents and what does it mean for machine agents to be part of a "command"? Billings' answer is that to be in effective command within a scope of responsibility, the supervisory agent (Billings, 1997, p. 39; also examine the role of the Flight Director in space mission control as in e.g. Murray & Cox, 1989):

• Must be involved
• Must be informed
• Must be able to monitor the automation or other subordinate agents
• Must be able to track the intent of the other agents in the system

The automated systems' and other subordinate agents' activities therefore must be comprehensible and predictable.

Billings' analysis provides us with a way to understand the problem-holder aspects of human roles in systems. People are problem-holders to the degree that they are responsible for the consequences of decisions and actions with respect to achieving various goals (see Tetlock, 1999; Lerner & Tetlock, 1999; Sharpe, 2004; Woods, 2004, for discussions of accountability). But Billings goes further, telling us that the problem-holder role in JCSs emphasizes the linkage of authority and responsibility—the person who holds responsibility for a problem is one who has some scope of authority to resolve the situation (see also Rochlin, 1999 and Weick et al., 1999 for additional observations that taking responsibility depends on having authority to act on the processes that influence outcomes).

Responsibility enters discussions of work because cognitive systems are goal-directed and because fields of practice always involve multiple interacting and sometimes conflicting goals. This leads us to several basic results on JCSs at work:

• Understanding a JCS means understanding the complexities of how goals interact and how specific situations give rise to conflicts among goals (see Woods, et al., 1994; Cook et al., 1998; goal-means analyses are important tools in CSE to the degree that they help chart how goals interact and how they conflict in different situations).

• Responsibility for achieving different goals and sub-goals is divided over multiple groups and roles; as a result, all JCSs are partially cooperative (all groups trying to achieve a common overarching goal) and partially competitive (the actions taken to achieve the goals one is responsible for may also undermine or squeeze the ability of others to achieve their goals).

• Dividing responsibility and authority creates double binds, which undermine performance of JCSs.

• The potential for goals to conflict creates fundamental tradeoffs and dilemmas that must be balanced in work at the sharp end of JCSs (see pp. 106-108; Woods et al., 1994, Chapter 4).

Quite surprisingly, issues on responsibility must be included in observing and modeling JCSs at work, and designing JCSs that work. Following Billings, the role of machine agents is limited with respect to issues of responsibility (Woods, 2002):

Computers are not stakeholders in the processes in which they participate. Whatever the artifacts and however autonomous they are under some conditions, people create, operate, and modify these artifacts in human systems for human purposes.

Goal Conflicts

Multiple, simultaneously active goals are the rule, rather than the exception, for virtually all domains in which expertise is involved. Practitioners must cope with the presence of multiple goals, shifting between them, weighing them, choosing to pursue some rather than others, abandoning one, embracing another. Many of the goals encountered in practice are implicit and unstated. In specific situations, goals often conflict; these conflicts are balanced through the actions of practitioners. Sometimes these conflicts are easily resolved in favor of one or another goal, sometimes they are not. Sometimes the conflicts are direct and irreducible, for example, when achieving one goal necessarily precludes achieving another one. But there are many intermediate situations, where several goals may be partially satisfied simultaneously. Multiple interacting goals produce *tradeoffs* and *dilemmas*. Resolving these tradeoffs and dilemmas takes place under time pressure and in the face of uncertainty. While some dilemmas arise from demands inherent in the process to be managed and controlled, organizations also constrain and pressure practice in ways that create or intensify dilemmas and influence the means available to resolve dilemmas (Figure 2). Any adequate analysis of a field of practice requires explicit description of the interacting goals, how they contribute to tradeoffs and dilemmas in particular situations, and how practitioners can handle them.

Part of the processes of cross-adaptation between blunt and sharp ends of a field of practice is the system of accountability—human decision making always occurs in a context of expectations in which one may be called to give accounts for those decisions to different parties (Tetlock, 1999; Lerner & Tetlock, 1999). Different systems of accountability can exacerbate or help resolve goal conflicts and dilemmas (Dekker, 2004; Woods, 2005a). The research results indicate—when organizations' systems of accountability create *authority-responsibility double binds*, they impose new complexities and dilemmas that undermine practice at the sharp end. Authority-responsibility double binds occur when a party has responsibility in that others may impose sanctions on that party following outcomes, yet that party no longer has sufficient authority to influence or control the processes that lead to outcomes.

For example, after the Three Mile Island accident, utility managers were encouraged by the Nuclear Regulatory Commission to develop detailed and comprehensive work procedures. Hirschhorn (1993) studied the impact when management at a particular nuclear power plant instituted a policy of verbatim compliance with all written procedures. This development occurred in a regulatory climate which believed that absolute adherence to procedures was the means to achieve safe operations and avoid "human error." However, for the people at the sharp end of the system who actually did things, strictly following the procedures posed great difficulties because:

(a) The procedures were inevitably incomplete and sometimes contradictory—brittleness.
(b) Novel circumstances arose that were not anticipated in the work procedures—demands for resilience. As a result, sometimes success could not be obtained if one only followed the procedure (Woods, 2006a).

Because accountability standards demanded strict adherence to procedures in this organization, the policy created a "double bind:" in some situations, if the operators followed the standard procedures strictly, the job would not be accomplished adequately; if they always waited for formal permission to deviate from standard procedures, throughput and productivity would be degraded substantially. If operators deviated and it later turned out that there was a problem with what they did (e.g., they did not adapt adequately), their actions could create re-work, safety, or economic problems. The double bind arises because the workers are held responsible for the outcome (the poor job, the lost productivity, or the erroneous adaptation); yet they did not have authority for the work practices because they were expected to comply exactly with the written procedures. As Hirschhorn put it:

They had much responsibility, indeed as licensed professionals many could be personally fined for errors, but were uncertain of their authority. What freedom of action did they have, what were they responsible for? This gap between responsibility and authority meant that operators and their supervisors felt accountable for events and actions they could neither influence nor control (1993, p. 140).

The double bind hinders the ability of practitioners to balance the tradeoff over the simultaneous risks of under- or over-adaptation to anomalies or disruptions (see p. 66). Consider the following situation from Cook, Woods & McDonald (1991). During a formal session where anesthesiologists gathered to discuss operating room cases that had not gone well, one case concerned decisions made during a Caesarean-section that had turned into an emergency. Recognizing that the case was just one instance of a deeper dilemma about how to balance the different goals and risks, one of the senior attending anesthesiologists asked the head of the unit what was the policy of the institution on how they should approach such tradeoff situations in emergency Caesarean-sections. Without pause the administrator (who was also an experienced practitioner) replied, "Our policy is to do the right thing."

This seemingly curious response, in fact, sums up the basic dilemmas inherent in the role of practitioner at the sharp end of complex systems. They are a problem-holder who is responsible for getting the job done successfully, regardless. Yet the organization may not support or may even hinder the practitioner from carrying out the role for responsible action in the face of irreducible uncertainty. In part, this lack of support between blunt and sharp ends of systems occurs because of the gap between distant images of work and actual work practices (Hollnagel et al., 2006). Distant images of work easily fall prey to over-simplification tendencies, missing the demands that shape expertise and create vulnerabilities for failure (Woods, et al., 1994; Woods & Cook, 2002). CSE arose in order to provide techniques for understanding the actual nature of practice (the core values of authentic observation and abstraction of generic patterns in work; see Figure 1). In part, the blunt end of the system fails to assist practitioners in handling dilemmas in advance because the organization seeks to cover its own responsibility to meet goals in relation to other stakeholders (Woods et al., 1994; Sharpe, 2004; Woods, 2004; Brown, 2005a; 2005b).

The curious response also captures the role of people to provide resilience, since all plans, procedures and algorithms are brittle (Hollnagel et al., 2006). It is impossible to comprehensively list all possible situations and encode all appropriate responses because the world is too complex and fluid (Law of Requisite Variety, and cf., e.g., Suchman, 1987; Woods et al., 1990 on the limits of procedures to match variation in the world). Thus the person in the situation is required to account for potentially unique factors in order to match or adapt algorithms and routines—whether embodied in computers, procedures, plans, or skills—to the actual situation at hand. This connects to the need to check if one is at risk to solve the wrong problem when one deploys an ostensibly related algorithm (the error of the third kind; Mitroff, 1974; see Chapter 8 on how this plays out in anomaly response). The Global Hawk mishap (pp. 113-115) illustrates the breakdown of literal-minded algorithms when there is no ability to test for whether the routine to be deployed actually fits the situation at hand. The example of space mission control center in anomaly response (pp. 69-70) illustrates a JCS which had adapted, given the high potential for surprise in space missions, to develop effective mechanisms to test the match of algorithms to potentially varying situations (see Patterson et al., 1999; Patterson & Woods, 2001).

Adapting to Double Binds

Practitioners adapt to constraints and pressures. They generally respond in one of two basic ways to double binds. When one party has responsibility in that they can experience sanctions based on outcomes, but lacks effective authority to influence the outcome sufficiently, one response is to try to pass responsibility back to others as well. This is sometimes referred to as a learned helplessness response as people distance themselves from both responsibility and authority. For example, by narrowly following rules, even when they are inappropriate, workers can reject responsibility when they do not possess appropriate authority (see this at work in the response of some practitioners in the authority-responsibility double bind created by the introduction of an expert system intended to de-skill and control operators in Roth et al., 1987).

Because practitioners are committed to getting the job done in the face of omnipresent organizational hurdles and other difficulties and gaps (Cook et al., 2000), a second response is more common—development of a covert work system. Practitioners develop one work system to get the job done given the inherent demands, tradeoffs, uncertainties and constraints imposed, while they appear to carry out another work system through formal documentation and other means to meet the expectations for providing accounts to other stakeholders at the blunt end of the organization (Hirschhorn, 1993; Bourrier, 1999).

Accountability systems can demand that practitioners give accounts based on adherence to procedures based on oversimplified distant views of the actual demands of work. Since procedures can never be a complete, consistent and coherent account of the skills and judgments required in practice, this expands the gap between the organization's image of work and the actual nature of work at the sharp end (Rochlin, 1999; Weick et al., 1999; Hollnagel et al., 2006). More effective organizations recognize that there will always be some gap between distant images of work and actual work practices, that there is a need to monitor for where and how this gap arises. For these organizations the gap represents an important source of information for learning and change to achieve better coordination between different levels of control (again, this coordination breaks down in the form of under- or over-adaptation to disruptions; p. 66).

Poor systems of accountability also degrade coordination across the groups that can make up a JCS. For example consider the current state of health care, which due to advances in capability and to pressures for efficiency (faster, better, cheaper pressures), requires more and more sophisticated forms of coordinated activity in order to achieve continuity of care (e.g., Patterson et al., 2002; 2004). The question then becomes—how do systems of accountability affect collaborative work and coordinated activity?

In order to carry out joint, interdependent activity, research has shown that the parties involved enter into a "Basic Compact," i.e., an agreement (often tacit) to facilitate coordination, work toward shared goals, and prevent the team's breakdown (Klein, Feltovich et al., 2005). One aspect of the Basic Compact is the commitment to some degree of aligning multiple goals. Typically, this entails one or more participants relaxing some shorter-term, more local goals in order to permit more global and longer-term goals to be addressed (Ostrom, 1990; Woods, 2006a).

Studies of cooperative exchange indicate that the basic compact is a form of positive reciprocity because overall results on critical goals such as patient safety or continuity of patient care depend on the joint impact of the actions and decisions of multiple parties over time. As one researcher described the complex reciprocal dependency, "Person 1 shows 'trust' for person 2 by taking an action that gives up some amount of immediate benefit in return for a longer run benefit for both, but in doing so person 1 relies on person 2 to 'reciprocate' in the future by taking an action that gives up some benefit in order to make both persons better off than they were at the starting point." (McCabe, 2003, p. 148).

Accountability systems based on sanctions and blame are one factor that can increase defensiveness and reduce commitment to this Basic Compact that underlies coordinated activity (Brown, 2005a; 2005b). If one party can experience high negative sanctions depending on outcome, they are less likely to relax their local goals as it exposes them to high risks when being called to account after outcome is known. Under threat from the system of accountability, each party defends their local goals rather than working toward aligning sub-goals to better achieve more global ends.

A second aspect of the Basic Compact is that all parties are expected to bear their portion of the responsibility to establish and sustain common ground and to repair it as needed (for a case of grounding see pp. 91-92; see also Clark & Brennan, 1998; Klein, Feltovich et al., 2005). Breakdowns in these processes are illustrated in the health care cases captured by Patterson et al., (2004) and Brown (2005a). Common ground refers to a *process* of communicating, testing, updating, tailoring, and repairing mutual understandings (and this is much more than a case of each party having the same knowledge, data, and goals). Detecting and correcting any loss of common ground that might disrupt the joint activity requires a significant investment of effort to track other groups and connect one's activities to their activities and needs (related to the workload economy of the JCS). The value of the Basic Compact is this shared willingness to invest energy and accommodate others, rather than just performing alone, walled off inside one's narrow scope and sub-goal.

In general, achieving coordination requires continuing investment and renewal, as parties to joint activity invest in those things that promote the compact and counteract those factors that could degrade it. These results point out how the blunt end of systems, though factors such as imposing accountability systems based on threats of high negative sanctions, can reduce the investment of effort in establishing, maintaining and repairing common ground, and can inadvertently encourage fragmented activities that should have been smoothly interconnected. Similarly, when the blunt end decomposition tactics to manage highly coupled processes compete with high demands for coordination to meet pressure for timely, efficient but very high performance, they can reduce investment in the functions that support coordinated and collaborative activities. In both cases, over-simplification tendencies at the blunt end increase the gap between the distant image of work and actual work; see Table 4).

Trust in coordinated activity occurs when all parties are reasonably confident that they and others will carry out their responsibilities within the Basic Compact (Ostrom, 2003). One danger of poor accountability systems is a breakdown in this form of trust and therefore the degradation of coordination across interconnected systems that is required for success in areas such as health care delivery.

LITERAL-MINDED AGENTS

The Global Hawk mishap at the opening of Chapter 10 is particularly important because it reveals general patterns about literal-minded agents first noted by Wiener (1950). The first pattern to note concerns the behavior of the UAV as a cognitive system—the automation did the right thing [given its model of the world], when it was in a different world! Second, the human role was to close the context gap between the automation's model of the world and the actual situation in the world.

Literal-mindedness creates the risk that a system can't tell if its model of the world is the world it is actually in (Wiener, 1950). As a result, the system will do the right thing [in the sense that the actions are appropriate given its model of the world], when it is in a different world [producing quite unintended and potentially harmful effects]. This pattern underlies all of the coordination breakdowns between people and automation noted in Chapter 10 (e.g., the list on p. 119).

How did this gap (the context gap) arise between the situation assumed in the model and the actual situation in the world? In the Global Hawk mishap, a software problem in the form of a hidden dependency was the proximal trigger. If one examines the broader set of observations on coordination breakdowns between people and automation, this is only one of a variety of complicating factors which can give rise to a context gap (Woods et al., 1990; Woods & Sarter, 2000).

In the Global Hawk mishap, did the different responsible human agents have the resources or support for their role to close or repair the context gap (why didn't they stop the automation)? The case provides a particularly vivid example of how design fails to support the changes in roles following the introduction of new levels of autonomy and authority for machine agents. If one examines the set of cases of coordination breakdowns between people and automation, the human contribution is that of failing to close the context gap.

The context gap is the need to test whether the situation assumed in the model underlying the algorithm to be deployed matches the actual situation in the world. Monitoring this gap is fundamental to avoiding the error of the third kind (solving the wrong problem), and to the demand for revision and re-framing in JCSs as discussed in Chapters 8 and 9.

These results point to a general pattern in technology change and JCSs at work: When a field of practice is about to experience an expanding role for automata, we can predict quite confidently that practitioners and organizations will adapt to develop means to *align, monitor, and repair the context gap* between the automata's model of the world and the world. Limits to their ability to carry out these functions will mark potential paths to failure—breakdowns in resilience. In addition, we note that developers' beliefs about the relationship of people and automation in complex and high consequence systems (substitution myth and other over-simplifications) lead designers to miss the need to provide this support and even to deny that such a role exists (but see Roth et al., 1987, Smith et al., 1997 and other studies of the brittleness of automata).

Now people, as a cognitive system, also are vulnerable to being trapped in literal-mindedness, where they correctly deploy a routine given their model of the world, but are in fact facing a different situation (Weick et al., 1999). When anomaly response breaks down, it is often associated with an inability to revise plans and assessments as new evidence arrives and as situations? change (pp. 74-75). An extreme form of people acting literal-minded happens in cases of fixation (pp. 76-77). As the Law of Demands (p. 19) points out, the vulnerability is a general one for any JCS at work.

But in observing work, we usually see practitioners probing and testing whether the routine, plan or assessment fits the actual situation they are facing (e.g., how mission control, as a JCS, works). This points to a fundamental difference between people and automata—people have the capability to repair the context gap for themselves (to some degree), but more powerfully, for and through others as part of a process of cross-checking and broadening checks across different perspectives.

Norbert's Contrast

Fitts' List—a comparison of things people do well and things computers do well—is one oversimplification about the relationship of people and automata at work (Hoffman et al., 2002). Almost as soon as Fitts' List was circulated, it was profoundly criticized as falling prey to a false opposition between people and computers. But what is the alternative conceptualization? Norbert Weiner's (1950) warning about the dangers of literal-minded agents provided the answer.

Norbert's Contrast
Artificial agents are literal minded and disconnected from the world while human agents are context sensitive and have a stake in outcomes.

The key is that people and computer automata start from opposite points—the former as context-bound agent and the latter as literal-minded agent—and tend to fall back or default to those points without the continued investment of effort and energy from outside. Automata start literal → developers exert effort and inventiveness to move these computer systems to be more → adaptive, situated, contextualized → but there are always limits in this process requiring human mediation to maintain or repair the link between model and actual situation (the fundamental potential for surprise that drives the need for resilience). On the other hand, people start contextualized → developers exert effort and inventiveness to move these human systems toward → more abstract, more encompassing models for effective analysis/action and away from local, narrow, surface models.

Norbert Weiner's (1950) original analysis revealed that literal-minded agents are paradoxically unpredictable because they are not sensitive to context. The now classic clumsy automation quotes from studies of cockpit automation—"What's it doing now? Why is it doing that? What will it do next?" (p. 119 and begun ironically by Earl Wiener, 1989)—arise from the gap between the human flight crew as context sensitive agents grounded in the world, and the literal mindedness of the computer systems that fly the aircraft.

The computer starts from and defaults back to the position of a literal-minded agent. Being literal-minded, a **computer can't tell if its model of the world is the world it is in**. This is a by-product of the limits of any model to capture the full range of factors and variations in the world. A model or representation, as an abstraction, corresponds to the referent processes in the world only in some ways. Good models capture the essential and leave out the irrelevant; the catch is that knowing what is essential and irrelevant depends on the goal and task context (Norman, 1993).

As Ackoff (1979, p. 97) put it,

The optimal solution of a model is not an optimal solution of a problem
unless the model is a perfect representation of the problem,
which it never is.

This "catch" on models captures the need to re-visit the connection between the model being deployed in the form of algorithms, plans, procedures, routines or skills and the actual conditions being faced, given change and the changing flow of evidence. This is the context gap that needs to be closed, as we saw in previous discussions of revising assessments, reframing conceptualizations, modifying plans in progress. The context gap is vividly illustrated by the examples in Figure 9 and the NASA memo about the computer alarms in Apollo on p. 88. In these cases, the same data or transition has many interpretations depending on a host of contextual factors.

It is up to people situated in context to ground computer processing in the world, given that particular situations can arise to challenge the boundary conditions of the model behind the algorithm (the potential for surprise). For people in context there is an open future, while literal-minded agents are stuck within the walls of the model underlying their operation. A practical example of the difficulty arises in the issue of anticipating limits to traversability as a robot encounters obstacles and changes in its environment. These judgments inevitably spill over the boundaries of the capabilities of the robotic platform and involve the people as handlers of a valued asset, and as responsible problem-holders confronting tradeoffs.

Closing the context gap is about knowing and testing what "rules" apply in what kind of situation. The key is to determine the kind of situation faced, and to recognize how situations can change. People can be sensitive to cues that enable them to switch rules or routines as they test whether the situation is different from what was originally construed. And despite not always performing well at this function (getting stuck in one view; pp. 76-77 and 104-105), people provide the only model of competence at re-framing or re-conceptualizing that we can study for clues about what contributes to expert performance and how to support it.

Automata (or any system that follows an algorithm, procedure, plan or skill mindlessly) need an outside perspective to align and repair the context gap. Thus, following an increase in automation, there is a predictable new risk or vulnerability to be avoided. The failure vulnerability is—the automation will do the right thing [given its model of the world] when it is actually in a different world, producing or contributing to a failure (Leveson, 2001; see the Global Hawk case pp. 113-115). To reduce this vulnerability requires, first, analyses of the brittleness of the automata. Second, it requires analyses of the sources of resilience in order to determine how and how well people are supported in their roles to align and repair the context gap (Hollnagel et al., 2006).

On the other hand, people as context-bound agents adapt to what they experience in the particular. This is a local perspective that is open to the potential for unbounded particularity, despite regularities. This perspective is important in the sharp end of practice as people bound to a local situation pursue their goals, with their knowledge, given their mindset (Woods et al., 1994, Chapter 4). The danger is that the local agent responds too narrowly to the situation in front of them, given that they can call to mind only a limited set of the relevant material and entertain only a limited set of views of the situation, missing the side effects of their actions and decisions at broader scales and time frames (Woods & Shattuck, 2000).

But, as learning agents, some people also participate in developing broader abstractions/models that, within bounds, provide a larger perspective on effective strategies (moving away from their context bound base point). Without continued effort, people tend to fall back to strategies that are dominated by local factors and context (Woods, et al., 1994, Chapter 4).

People create procedures and automata to connect the abstractions developed back to guided actions in the world. But these abstractions and these developers form a distant perspective on work in particular contexts. This perspective then supplies automata and plans, but is trapped in the walls formed by the underlying model (its assumptions and boundary conditions). It takes effort to step outside those walls in order to gain perspective on changes in the world to be managed.

With advances (developer effort), computer agents can be made more situated in the sense of taking more factors into account in their own processing. These automata still function well only within artificially bounded arenas (although the size and position of these arenas grow and shift as people learn to produce new capabilities). Literal-minded automata are always limited by the brittleness problem—in other words, however capable, developer effort stretches automata away from their literal-minded origin, but without chronic effort to ground these systems in context, the risk of literal-mindedness re-emerges.

This work to close the context gap tries to keep the environment where the automata are placed in alignment with the assumptions underlying the (see the discussion of clumsy automation in Chapter 10). However, the effort to ground literal-minded automata in context often occurs in the background, hidden. Nevertheless, it takes continued effort to maintain this match of model-environment in the face of the eroding forces of variability and change inherent in our physical world (the potential for surprise inherent in the Law of Requisite Variety). And, as the envelope of competence of automata shifts, the resulting new capability will afford adaptation by stakeholders in the pursuit of goals. This creates new contexts of use that may challenge the boundary conditions of the underlying algorithms (as captured in the Law of Stretched Systems, p. 18, and as has been experienced in software failures that arose from use migration and requirements change).

Thus, Norbert's Contrast specifies two complementary vectors:

(1) The capabilities of automata grow as (some) people learn to create and instantiate new algorithms, plans and routines;

(2) In parallel, people at the sharp end adapt to the brittleness of literal-minded agencies to monitor, align and repair the context gap omnipresent in a changing uncertain world (and the change results, in part, from how people adapt to exploit the new capabilities).

The contrast captures a basic tradeoff that must be balanced in the design of any JCS because of the basic constraint of bounded rationality of finite resources and irreducible uncertainty (p. 2). Literal-minded and context-bound agents represent different responses to the tradeoff, each with different vulnerabilities. Norbert's Contrast points out that literal, disconnected agents can be powerful help for context bound agents, but the literal-minded agents in turn need help to be grounded in context (Clancey, 1997). Coordinating or complementing different kinds of cognitive agents (literal-minded and context-bound) in a joint cognitive system is the best strategy for handling the multiple demands for performance in a finite-resource, uncertain, and changing universe. This complementarity essentially recapitulates the epigraph on p. 2 and Figure 1 on p. 5. In many places and in many ways, this balancing act between the abstract and the particular defines the basic heuristic strategy of CSE—insight will be found in shifts and contrasts between the particular and the universal.

DIRECTIONS FOR DESIGNING JOINT COGNITIVE SYSTEMS THAT INCLUDE ROBOTIC PLATFORMS

When one applies findings about JCSs at work to envisioning and guiding the effects of technological change, such as in the case of human-robot coordination, what results?

Reverberations of New Robotic Technologies

The substitution myth has produced a fixation on autonomy as the goal for the design of agents and artifacts for work. Instead, a JCS approach recognizes that technological powers are important relative to how they afford adaptation to support human systems achieve human goals.

When we break the fixation on autonomy, we notice that many technological advances can be viewed as means for perception at a distance or action at a distance (see *JCS-Foundations* on amplification, pp. 25-33; Flach et al., 2004). In these cases, technology extends our perception through sensors and scopes or extends our activities in terms of the sequences, precision, or forces we exert indirectly on the world (e.g., one act triggers a sequence of activities, or one activity is translated into the component physical actions needed to accomplish intent, as in modern aircraft controls).

New capabilities for robotic systems are a major step forward within this tradition of coupling people to scenes at a distance (Woods et al., 2004). People in various roles participate in the JCS as problem-holders, i.e., those people and groups responsible for achieving goals. Robotic systems provide a new form of perception-action coupling: at-a-distance coupling, which connects the purposes of the problem-holders to assessments and activities in physically remote environments.

New power (more sophisticated perception-action coupling as robotic systems are endowed with sufficient capability to move on their own beyond tele-operation only) does not remove the human from the scene, but ironically couples them in a way that is paradoxically intimate, though physically removed (or mediated). Robotic systems are fascinating because they provide human problem-holders with higher order means to achieve their goals by *projecting human intent into the world*. Ultimately, robotic capabilities represent new powers for human problem- and stakeholders to project intent at a distance. This is evident in the findings of Murphy in deploying robots in search and rescue field exercises, and in actual emergencies (Casper & Murphy, 2003; Burke et al., 2004; Murphy & Burke, 2005). She and her colleagues have found that when robotic platforms provide value, that value is relevant to an increasing set of problem-holders responsible for achieving different goals as part of the mission.

> Human-robot interaction should not be thought of in terms of how to control the robot, but rather as how a team of experts can exploit the robot as an active information source. Search and rescue is an information process which involves both a hierarchy (search team, task force, incident commander) that filters information according to task and distributed users (search, medical, hazmat, structural collapse, safety) who use the same data from a robot in different ways (Murphy & Burke, 2005, p. 3)

> Many of the "information consumers" may be geographically distributed and have different tasks. ... The distributed users have competing goals, yet must work as a team. ... The lesson derived from additional studies focusing on how these distributed teams can cooperate in Robot-assisted medical reachback show that they use shared visual information to build shared mental models and facilitate team coordination. (Murphy & Burke, 2005, p. 3)

The requirement for new forms of coordination that meet pressing needs and that are enabled by new technology is striking in this synthesis of field results. People become involved and interconnected in new roles, across roles, and to demands and purposes in new ways. The demands for coordination arise in part due to the inherent demands emergency response places on any JCS:

> ... our experience [is] that every response is totally different and causes unforeseen problems or opportunities. We have never gone to an actual response and used the equipment the way we thought we would. (Murphy & Burke, 2005, p. 4)

To paraphrase a common expression from military decision making:

"No plan survives contact with a disaster-in-the-making."

Naturalistic decision-making requires adapting plans to handle the unique evolution of particular threats (see Klein, 1999; Klein et al., 2003; Smith et al., 2004 or, for a particular case, see how mission control functions as a JCS re-examine the episode on pp. 69-70). Not all fields of practice have to confront domains where the potential for surprise is so high, but all JCSs are adapted to the potential for surprise in their fields of practice (the fundamental place of the Law of Requisite Variety and its implications for work as control and adaptation in *JCS-Foundations*, pp. 40-47, Chapter 7). The *potential for surprise* is related to the next anomaly or event that practitioners will experience and how that next event will challenge pre-developed plans and algorithms in smaller or larger ways. To assess the potential for surprise in a work domain one considers how the above generalization applies to that particular setting—*how do plans survive or fail to survive contact with events*, and one searches for the kinds of situations and factors that challenge the textbook envelope in a field of practice (Woods et al., 1990; Woods & Shattuck, 2000; Woods, 2006a).

All JCSs are adapted to the potential for surprise, not simply to frequently encountered demands (revisit the functional account of anomaly response in Chapter 8). Resilient systems are *prepared to be surprised* (Rochlin, 1999; Woods, 2005b). Note how this concept has already been reverberating through parts of this book. Discussion on how to study JCSs at work is centered on the concept that scientific observation is also based on methods that help observers to be prepared to be surprised (Chapters 4 and 5). The functional synthesis of anomaly response revealed the central role of the ability to re-conceptualize as situations change; this ability is based on processes that support being prepared to be surprised (see Chapter 8).

The results on robots in search and rescue are just one example of a work setting undergoing change that illustrates how the Law of Requisite Variety modulates JCSs in action. And this case of change illustrates how people at the blunt end of systems apply decomposition and substitution heuristics, which are over-simplifications about work at the sharp end (Table 4). When the blunt end is trapped in such oversimplifications, the organization will fail to support the sharp end as it adapts to cope with new forms of complexity, and as pressure to be *faster, better, cheaper* squeezes the dilemmas and demands present in practice (Woods, 2006a).

The basic asymmetry in coordinative competencies between people and automata, (Robin) Murphy's Law, and other findings from human-automation teamwork remind us of the limits of automata in coordinated activity (brittleness, literal-mindedness). Given the inherent potential for surprise in complex settings and limits of automata, there is another human role in the ensemble that must be planned for in designing JCSs that work. The *Robot Handler* role is responsible for managing the robotic capabilities *in situ* as a valued resource, and points to the knowledge, practice, and interfaces needed to manage the robotic resource in a physical environment. For example, Casper & Murphy (2003) found that, given the demands of search and rescue operations and the limits of current robotic systems, the search and rescue personnel functioned as problem-holders trying to characterize the search situation and achieve rescue goals, while the robot developers who brought the new devices to the field acted as the handlers who better understand robot capabilities and limits, and directed the robot's capabilities.

In complex settings, difficulties cascade and demands escalate, which will challenge robotic systems' ability to compensate and demand coordination between people and robots (Woods & Patterson, 2000). Inevitably, robot capabilities will exhibit brittleness as situations develop beyond their boundary conditions (e.g., Roth et al., 1987; Guerlain et al., 1996; Smith et al., 1997). Together, these represent challenges to the adaptive power or resilience of the human-robot ensemble (as the fit between the robotic systems and the mix of constraints in the world in question is always partial).

Coordinating these roles within the physical limits of robotic systems creates critical guiding questions for assessing coordination: How will human team members recognize the approach to brittle boundaries and intervene effectively (e.g., bumpless transfer of control)? Inevitably, autonomous resources will be lost, or will fail. How will the team dynamically reconfigure or gracefully degrade as assets are lost? One example of a function that tests coordination across these human and robotic roles is judgments of traversability or climbability of terrain in context. How does the JCS recognize when terrain challenges the boundary conditions of the robotic platform's mobility (given the risk of not accomplishing goals) and potential for recovery if the platform gets stuck or consumes too much time or energy (Murphy, 2004).

To achieve new levels of coordination (Chapter 10), past research has shown that increases in the level of autonomy and authority of automata require an increase in *observability*—the levels and kinds of feedback between agents about their current, but especially *future*, activities as system state varies (Christoffersen & Woods, 2002). Field studies, incidents and simulation results all reinforce this as a basic finding. When this relationship is ignored, coordination surprises occur between agents.

The research challenge is to define the forms of feedback needed for seeing into future activities and contingencies (Roth et al., 1997; Woods & Sarter, 2000). Critical to the new forms of feedback are representations of automation activity that capture events, are sensitive to future developments, and integrate data into higher order patterns—not display of base data elements, current process state or automation configuration. For example, as robotic systems have the capability to follow plans and to shift to a new plan as the situation changes (e.g., current UAVs), how will human supervisors monitor robot plan selection and plan following, recognize disrupting events, and modify plans?

The studies also have shown that increases in the level of autonomy and authority of automata require mechanisms to re-direct automated systems as resources–*directability* (Klein et al., 2004). This refers to the capacity for deliberately assessing and modifying the actions of the other parties in a joint activity as conditions and priorities change. Effective coordination requires adequate responsiveness of participants to the influence of others as the activity unfolds. This means that development of machine agents also needs to provide means for controlling aspects of agent autonomy in a fashion that can be both dynamically specified and easily understood (Myers & Morley, 2003). One promising direction that is being explored is that of using "policies" as a means to dynamically regulate the behavior of a system without changing code or requiring the cooperation of the components being governed (Bradshaw et al., 2004, Feltovich, Bradshaw et al., 2004). Through policy, people can precisely express bounds on autonomous behavior in a way that is consistent with their appraisal of an agent's competence in a given context. The machine agent's behavior becomes more predictable with respect to the actions controlled by policy. Moreover, the ability to change policies dynamically means that agents performing poorly can be adjusted and managed to contribute still to achieving goals. It is hoped that developing this form of human-automata coordination will meet the criteria for inter-predictability and sharing of intent in effective joint activity as well (Billings, 1997; Shattuck & Woods, 2000; Klein et al., 2004).

To summarize: Coordinating People and Computers in JCSs at Work

The cases and concepts in Chapters 10 and 11 reveal the necessity of the concept of a JCS and the limits of oversimplifications that focus exclusively on autonomy of machine agents.

*As we **participate** in roles as developers at the blunt end, the results from the sharp end remind us that, autonomy alone is not enough*

- *To support strategies for managing complexities in work,*
- *To anticipate the dynamics of cross-adaptive processes,*
- *To balance the tradeoffs and dilemmas that lie underneath the adaptations of work.*

Success comes from coordinating over the agents and functions in the JCS to be resilient in the face of the demands of fields of practice. Technological advances continue to feed new power to the search for what will support JCSs at work (affordances).

Chapter 12

Laws that Govern JCSs at Work

Now all scientific prediction consists in discovering in the data of the distant past and of the immediate past (which we incorrectly call the present), laws or formulae which apply also to the future, so that if we act in accordance with those laws our behavior will be appropriate to the future when it becomes the present.

Craik, 1947, p. 59

A TACTIC TO REDUCE THE MIS-ENGINEERING OF JOINT COGNITIVE SYSTEMS

Observation and studies of JCSs at work have built a body of results that describes how technology and organizational change transforms work in systems. Points of technology change push cycles of *transformation* and *adaptation*. The common observation is that technological change does not reduce workload or simplify the tasks of operators. Instead, the new capabilities it creates are exploited by people responsible for achieving various goals "to ask personnel to do more, do it faster, and do it in more complex ways" (Cordesman & Wagner, 1996, p. 25).

On the other hand, people in roles at the blunt end act as advocates for investment in and adoption of new technology. In these roles, people make claims about how these changes will affect work in the target field of practice. Claims about the future of practice if objects-to-be-realized are deployed represent *hypotheses* about the inter-relationship of people, technology, and work (Woods, 1998). Observations at points of technology change find that these hypotheses can be, and often are, quite wrong—a kind of second order automation surprise (Sarter et al., 1997). Envisioning the future of operations, given the dynamic and adaptive nature of work, is quite fragile.

What patterns emerge from observations of people engineering work? What patterns emerge from observing the gap between the actual effects following re-engineering of work and the previously made claims about how various advances-in-the-making would re-engineer work? Consistently, we observe over-simplifications (Table 4; Feltovich et al., 1997) that claim that the introduction of new technology and systems into a field of practice substitutes one agent for another—essentially, a claim that consistent computer capabilities substitute for

167

erratic human performance. These claims of an opposition of human and machine have come cloaked in different and sometimes quite sophisticated forms, yet underneath remains the core of this myth—capabilities of people versus capabilities of machines are an appropriate unit of analysis so that new technology (with the right capabilities) can be introduced as a simple substitution of machine components for people components—preserving the overall system though improving the results that can be obtained (Fitts' List). This over-simplification fallacy is so persistent it is best understood as a cultural myth (Woods & Tinapple, 1999).

The myth creates difficulties because it is wrong empirically—adding or expanding the machine's role changes the JCS, changing human roles, introducing new capabilities and complexities, triggering new rounds of adaptive cycles as human actors and stakeholders jostle in the pursuit of their goals. But moreover, the myth is unproductive as it

• locks us into cumbersome trial and error processes of development,
• blocks insight into the demands of work in context,
• misses how people in various roles and groups adapt to these demands, and
• channels energy away from innovative exploration of what would be useful ways to use the continually expanding power of technology for data pick up, data transmission/collection and handling, autonomous action, and connectivity at a distance.

How can we better calibrate and ground claims about the future of work to avoid past cycles where change produced clumsy technology and placed new demands on the people responsible to meet system goals? One possible tactic is to develop generalizations or laws that govern work by any cognitive agent or any set of cognitive agents from the base of empirical observations. The label "law" is used here in the sense of (a) a compact form of generalization, (b) patterns abstracted from diverse observation, and (c) a means to help handle future situations to be encountered, as captured in the quote that opens this chapter from Craik in his pursuit of theories of control that include the operator (cf., also, criteria on p. 56).

In this closing chapter, we synthesize a set of such laws as tentative generalizations. The statements are law-like in that they capture regularities about amplification of control in work (*JCS-Foundations*, p. 28). The tactic is a deliberate play off Conant's (1976) attempt to specify "Laws of Information that Govern Systems."

The set is intended

• To help jump start new opportunities to observe JCSs at work in various "natural" laboratories
• To stimulate others to take up the challenge of abstracting general patterns from observations across different work settings and through different gathering techniques
• To provide a principled base for collaboration with other stakeholders in the process of innovating new support for work

This set of laws is built on a foundation of mutual adaptation as people adapt to achieve goals given the demands of work settings and the affordances of the artifacts available (pp. 7-8). This means that the processes to be modeled are cross-adaptive. Agents' activities are understandable only in relationship to the properties of the environment within which they function; an environment is understood in terms of what it demands and affords to actors in that world. Each is only understandable in relationship to the other so that one is a necessary background or frame to help bring the other target into focus as foreground.

FIVE FAMILIES OF FIRST PRINCIPLES OR LAWS

This section collects various generalizations and repeating patterns as a set of first principles about how JCSs work. Many of these have appeared in the text already as introductions to or conclusions from stories of JCSs at work. Some have not appeared explicitly, but their role can be seen in the stories of coordination, resilience, affordance (or their opposites). The laws are grouped into five families, each capturing a surprising insight about the nature of work (the set of laws and related resources also can be examined at url: http://csel.eng.ohio-state.edu/laws).

Overall, the driving force behind the laws is the way joint cognitive systems adapt to variations, and the potential for surprise in the worlds of work. Hence, the first family of laws is *Laws of Adaptation* or control. The surprise is that understanding work begins with understanding how it is adapted to cope with complexity—the foundational slogan for Cognitive Systems Engineering (Rasmussen & Lind, 1981; Woods, 1988; *JCS-Foundations*, p. 65 and Chapter 4). This family of laws starts with the original insights of cybernetics and control (Ashby, 1957; Wiener, 1948; Conant, 1976; see *JCS-Foundations*, Chapter 7). This subset of laws addresses the challenge of how JCSs can be resilient—the ability to anticipate and adapt to potential for surprise and error.

The second set concerns *Laws of Models*. When we consider how systems adapt to control in the face of variability, the surprise is that expertise is tuned to the future. This theme of anticipation runs through almost all of the stories of JCSs at work captured in this book (Hollnagel, 2002). But how can expertise be based on anticipation, i.e., future trajectories and contingencies, when the data available are only about the past, and the models that could be run out to predict the future can be overwhelmed by uncertainty and change?

As was noted in the original work of cybernetics (Conant & Ashby, 1970), the basis for control, adaptation and anticipation comes from models. Effective control is a model of the processes managed and the variations confronted (*JCS-Foundations*, pp. 40-41, 186). The Fluency law is essentially a restatement of Conant & Ashby's original insight.

Thus, this family of laws is concerned with how we understand and represent the processes we control and the agents with which we interact. A good model focuses on the relevant and leaves out the irrelevant. The limit is that different models make different distinctions about what is relevant, and represent those distinctions differently, changing the work needed to use that model/representation to solve problems (Norman, 1993; Zhang & Norman, 1994). As situations change or new evidence comes in, we also face the challenge of shifting from one model to another or deciding which of several models is relevant as a guide to handle the situation at hand. For example, as disturbances from a fault cascade, the JCS for anomaly response may need to shift from a simpler to a more sophisticated model of the process being controlled.

Laws of Coordination form the third set of laws and address how work is distributed and synchronized over multiple agents and artifacts (Hutchins, 1995a; Klein et al., 2003). The surprise here is the fact that work always occurs in the context of multiple parties and interests as moments of private cognition punctuate flows of interaction and coordination. The idea that cognition is fundamentally social and interactive, not private, radically shifts the basis for analyzing and designing work, and forces a deep reconsideration of the relationship between people and computers—the JCS. This insight means that CSE approaches work "middle-out" (see the discussion on systems concepts in CSE, p. 7). It begins with coordinated activity to meet demands as a middle layer that provides a context for moments of private cognition (e.g., how attentional processes function in multi-agent interactions as one follows another's focus), and, in turn, is influenced by constraints imposed from "blunt end" organizational processes (e.g., how threat of sanctions after outcome is known shapes adaptation in practice at the "sharp end").

Quite surprisingly, *Laws of Responsibility* are the fourth family, driving home the point that in cognition at work, whatever the artifacts, and however autonomous they are under some conditions, people create, operate, and modify these artifacts in human systems for human purposes. People, and only people, can assume roles as problem-holders. Here we see how the fundamental potential for goals to conflict drives processes of work.

The fifth family addresses the relationship of people and computers/automata in work—though more broadly the issue concerns how people in context relate to all of the routines, rules, plans, procedures, and algorithms that we develop at a distance and instantiate in varying physical forms whether on paper, in skills, or in automata. The surprise here is the caution raised by Norbert Wiener (1950) warning us of the limits of literal-minded agents and the unique competencies of people to handle the tradeoffs and dilemmas of a changing, finite-resource, uncertain, conflicted world (and see the discussion of the psychologist fallacy pp. 52-53 on how distant views can be limited and over-confident). *Norbert's Contrast* goes behind our culture's fascination with increasing the power of the computer to remind us of how we actually expand success despite finite resources, irreducible uncertainty, conflicting goals, and omnipresent change.

One hallmark of the human species, along with language and thought, always has been skilled creation and use of tools. Artifacts are not tools in themselves (despite the tendency of technologists to label "kitchen sink" collections of features and capabilities as "tool kits"), but rather, artifacts become tools through skilled use in context. An icon for this basic fact about people and the artifacts they create is the image sequence that occurs as the ending of the opening sequence in Stanley Kubrick's movie *2001*—**Kubrick's Bone** (see Woods & Tinapple, 1999). One group of early hominids learns to use objects around them as tools; that discovery is marked by the visual symbol of a monolith. One, after having used an animal bone as a weapon, exults and throws the tool into the air; the image sequence then shifts as the object thrown comes down as a spacecraft when the movie jumps forward centuries. The image of Kubrick's Bone reminds us of another basic slogan in CSE, Don Norman's (1983) phrase, "things that make us smart" (or dumb). When we are tempted to attribute intelligence to the artifacts we fashion with each successive wave of new technology, we should instead recall this phrase used near the beginning of CSE: "The intelligence in decision support lies not in the machine (tool) itself but in the tool builder and the tool user." (Woods, 1986, p. 173).

LAWS THAT GOVERN JOINT COGNITIVE SYSTEMS AT WORK

1. Laws of Adaptation:
Driving Force: JCSs adapt to cope with complexity.
Challenge: How to be resilient given the varying potential for surprise in worlds of work.

Law of Requisite Variety
Only variety can destroy variety.

Context conditioned variability
Skill is the ability to adapt behavior in changing circumstances to pursue goals.

Law of Stretched Systems
Every system is stretched to operate at its capacity. As soon as there is some improvement, some new technology, we exploit it to achieve a new intensity and tempo of activity.

Law of Demands
What makes work difficult, to a first approximation, is likely to make work hard for any JCS regardless of the composition of human and/or machine agents.

Law of Fluency
"Well"-adapted cognitive work occurs with a facility that belies the difficulty of the demands resolved and the dilemmas balanced.

Potential for Surprise
All JCSs are adapted to the potential for surprise in their fields of practice—how do plans survive or fail to survive contact with events.

The potential for surprise is related to the next anomaly or event that practitioners will experience, and to how that next event may challenge pre-developed plans and algorithms in smaller or larger ways (Woods et al., 1990; Woods & Shattuck, 2000). Resilient systems are *prepared to be surprised*, i.e., set up to re-frame and re-conceptualize assessments of how well plans/models/automata fit particular situations to be handled (Woods, 2006a).

Commentary

Studies and models of JCSs always begin with examining how practice has adapted to handle variability. Of particular importance is the fact that events and situations can arise that challenge the boundary conditions on how practice has adapted—the potential for surprise. The promising direction for new designs that would support work is amplifying control in the face of variability in the world.

One should note how the word "surprise" recurs throughout the discussions of JCSs at work. Anomaly response and discovery in science are based on recognizing a surprising finding (Peirce, 1955). Being prepared to be surprised is critical to revise assessments to avoid being stuck in one view when the situation to be handled is different—resilience. Despite the deep roots for these concepts, the findings of CSE are quite surprising, relative to commonplace assumptions in our culture about the nature of work, what makes it difficult, and how technology affects work.

2. Laws of Models
Driving Force: The mystery of how expertise can be tuned to the future when the data available are about the past.
Challenge: How models are both essential and limited as we try to understand and represent the processes we control and the agents we interact with.

Avery's Wish (1998):
Usability Engineer:
   What *features* would you like to see in the new
   [2002 agent based, augmented, wearable, ...]? [fill in today's date and technology fads]
Practitioner:
   I yearn for some thing that shows me an image of what the system and the room are going to look like, ten minutes from now.

Law of Regulation
Regulation is a model of what is regulated relative to disturbances, or,
Every good controller must be a model of what it controls (Conant & Ashby, 1970)
(a) The structure, function, and dynamics of the processes regulated,
(b) The variability of the environmental disturbances and changes to be compensated for.

Ackoff's Catch
The optimal solution of a model is *not* an optimal solution of a problem unless the model is a perfect representation of the problem, which it never is (Ackoff, 1979). A good model focuses on the relevant and leaves out the irrelevant; the *catch* is being able to tell what is relevant or knowing when the situation demands a shift to a different model.

The Representation Effect
How a problem is represented changes the work needed to solve that problem, either improving or degrading performance, for any cognitive system or joint cognitive system (Zhang & Norman, 1994).

Law of Anticipation
Feedback, essential for control, is the property of being able to adjust future conduct by past performance (Wiener, 1964, p. 33), i.e., *expertise is tuned to the future*, while, paradoxically, data are about the past.

Law of Expectations
People's behavior depends not on what did happen, but on what *could* have happened, and especially what could happen *next*.

There is No Such Thing as a Cognitive Vacuum
When we [as designers] create, represent, or shape some experience for others, those people learn something [even if it is not what we intended]; form some model or explanation [even if it is not veridical]; extract and index some gist [even if it is quite different from what was relevant to us]; see some pattern [even if we encoded other relationships or no relationships at all]; pursue some goal [even if it is not the goal we most value]; balance tradeoffs [even if our idealizations about work push them out of our view].

Commentary

Anticipation is a constant in stories of JCSs at work (e.g., Klein et al., 1995; Klein, 1999). Being tuned to what could happen next and establishing a readiness to respond as situations develop and change is the hallmark of expertise, as well as a mystery.

To anticipate depends on expectations derived from models that capture the influences at work on the processes being managed. But models bring a catch, as they are always representations of the world that leave out some aspects and highlight other aspects. As a result, the need to test the fit of models to situations and the need for revision and re-focusing dominates JCSs. CSE has used this aspect of JCSs as a basic strategy by studying how to make models of processes tangible and observable in new representations, given the capabilities of new media (Norman, 1993; Woods, 1995a; Flach et al., 2003).

3. Laws of Collaboration
Driving Force: Work always occurs in the context of multiple parties and interests as moments of private cognition punctuate flows of interaction and coordination.
Challenge: How to coordinate, synchronize, and cross-check work over multiple agents and artifacts as situations change.

First Law Of Collaborative Systems
It's not collaboration, if either you do it all or I do it all.
Collaborative problem solving occurs when the agents coordinate activity *in the process of* solving the problem.

Second Law of Collaborative Systems
You can't collaborate with another agent if you assume they are incompetent.
Collaborative agents have access to partial, overlapping information and knowledge relevant to the problem at hand.

Coordination Costs, Continually
Achieving coordinated activity across agents consumes resources and requires ongoing investments and practice.

Computers are not situated
Computers can't tell if their model of the world is the world they are in.
Computers need people to align and repair the context gap.

The Collaborative Variation on the Law of Regulation
Every controller is a model of the other agents who influence the target processes and its environment and of those who coordinate their activities directly or indirectly to achieve control, i.e., a model of the activities, models, capabilities, and expectations of the other agents.

Mr. Weasley's Rule [15]
Never trust anything that can think for itself
if you can't see where it keeps its brain?
                                                Harry Potter

Commentary

A common trap that results from the pursuit of isolated autonomy is systems where "either you do it all or I do it all" (see the stories on pp. 113-116). Such designs fail to coordinate and collaborate in handling situations. For example, Mr. Weasley's rule is a succinct statement summarizing the lines of research on common ground in cooperative work and the failure of machine explanation to make intelligent computers team players (see pp. 91-92 and 132-134).

In addition, the corollary to the psychologist's fallacy (pp. 52-53) reminds us that success derives from coordinating multiple agents each with only partial information, knowledge and skills (e.g., the updating case, pp. 91-92). When we assume that "other agents have a weaker grasp ... than we do," design comes to focus on protecting our plans, algorithms, and automata from contamination by those other erratic people. But no agent is immune to the constraint of bounded rationality; so developers who ignore designing for coordination across agents have

---

[15] From *Harry Potter and The Chamber of Secrets*, 1999 (italics in original), p. 329.

fallen prey to oversimplifications and overconfidence tendencies that have resulted in the many automation surprises (Woods & Dekker, 2000; Roesler et al., 2001). The irony remains that the new powers of technology create new demands to design for coordinated and synchronized activity.

These are two examples of how new findings are questioning and overturning folk models about coordination and collaboration (Olson et al., 2001; Hutchins, 1995a; Heath & Luff, 1992). One omnipresent oversimplification is to mistakenly believe that providing the technical capability to connect to disparate parties and data sources—connectivity—also automatically provides the performance capability to synchronize and cross-check assessments and activities—coordination. There are other functions to be supported and investments to be made if JCSs are to demonstrate sustained high levels of coordination and collaboration (e.g., Klein, Feltovich et al., 2005).

4. Laws of Responsibility
Driving Force: Computers are not stakeholders in the processes in which they participate.
Challenge: There are always multiple relevant goals that can conflict in particular situations.

Computers are not stakeholders in the processes in which they participate. Whatever the artifacts and however autonomous they are under some conditions, people create, operate, and modify these artifacts in human systems for human purposes.

Computer artifacts are stand-ins for distant human groups and organizations (and their models of local conditions).

Conflicting Goals
Multiple but potentially conflicting goals apply to all systems.

Differential Responsibility
Different parts of a distributed system are differentially responsible for different subsets of goals that can interact and conflict.

Commentary
As Wiener (1950) & Billings (1997) emphasized when considering automata, only people can be held responsible for outcomes by other stakeholders. Hence it is important to consider how people fulfill roles as problem-holders, i.e., the human roles responsible for achieving mission goals, and the associated knowledge and experience needed to carry out that responsibility. As problem-holders, people participate in a reciprocating cycle of giving accounts and being called to account by other stakeholders for decisions and actions (Lerner & Tetlock, 1999; Tetlock, 1999).

The first constraint on responsibility in JCSs is that artifacts, however autonomous, are not and cannot be problem-holders. People may pass targets for automata to pursue and pass constraints for automata to respect, but this is not the same as being responsible for goals. The second constraint on JCSs given responsibility is that there are always multiple relevant goals at different levels, which can conflict in particular situations (which means that distributed systems are always in part competitive and part cooperative as in the tragedy of the commons; Ostrom, 1990). Hence, action always occurs relative to a background of dilemmas and tradeoffs (pp. 106-108). Usually implicitly, practice adapts to find ways to balance the tradeoffs (cf., the story on p. 153; Appendix B).

Issues of responsibility inevitably lead to discussions of trust in coordination of automata and people. While some forms of autonomous machines are necessary and uncontroversial elements of many systems (for example, no commercial jet aircraft could be built or operated today without advanced control automation), human managers and stakeholders (mission personnel), given their responsibility and accountability to achieve an organization's goals, may not trust adaptive autonomous systems, as they seem to introduce new complexities and are hard to monitor and manage (Billings, 1997). In part, this arises because developers have consistently misunderstood the boundary conditions for autonomous performance and underestimated the need for adaptability and coordination (Murphy, 2004; Woods et al., 2004). Freed et al. (2004) found that many new autonomous systems developed for NASA have not been flown on missions because they introduce unacceptable sources of uncontrolled risk for NASA managers given the criticality of spaceflight projects. Freed raises the question of what would be trusted autonomy since it would need to be based on the perceptions of stakeholders about the processes and organizations that determine how the software is maintained and improved, and the affordances the software offers to mission personnel that allow its behavior to be understood, managed, tested, directed, and communicated (misunderstandings relative to the above can also lead to misplaced trust).

5. Norbert's Contrast of People and Computers
Driving Force: People create and wield artifacts as tools, even autonomous machines, to amplify human control in the face of variability (not an opposition or substitution of machines for people).
Challenge: How to create success through distributed work to balance tradeoffs and dilemmas.

Bounded Rationality Syllogism
All cognitive systems are finite (people, machines, or combinations). All finite cognitive systems in uncertain changing situations are fallible. Therefore, machine cognitive systems (and joint systems across people and machines) are fallible.
The question, then, is not fallibility or finite resources of systems, but rather the development of strategies that handle the fundamental tradeoffs produced by the need to act in a finite, dynamic, conflicted, and uncertain world.

Joint Systems Maxim
People and computers are not separate and independent, but are

interwoven into a distributed system that performs work in context—a Joint Cognitive System. Success is created and sustained through the work accomplished by the joint cognitive system; it is not a given which can be degraded through human limits and error.

Norbert's Contrast
Artificial agents are literal-minded and disconnected from the world, while human agents are context-sensitive and have a stake in outcomes.
People and computers *start* from opposite points (one literal-minded and the other context-sensitive) and tend to *fall back* or default to those points without the continued investment of effort and energy from outside that system.

Poly-centric Balance
All systems are balancing distant plans with local adaptations to cope with the potential for surprise.

Commentary
To be effective in our universe of bounded rationality requires complementary processes in order to balance the risks across the different traps inherent in the dilemmas and tradeoffs of practice. The models captured in plans/procedures/algorithms/automata are both necessary to cope with this complexity and insufficient, given the potential for surprise. The potential for gaps in any algorithms/plans/procedures is basic (Suchman, 1987); thus, resilient systems are *prepared to be surprised*, that is, they are set up to re-frame and re-conceptualize assessments of how well models embodied in plans/procedures/algorithms/automata fit the actual situation to be handled (Woods, 2006a). To do this well depends on coordination across agents, which amplifies control through the contrast and combination of multiple perspectives. Different agents in different roles have different views, knowledge and stances about the situation in the world. In information analysis, for example, the results from different computer algorithms define different perspectives that the human monitor can shift across or contrast with each other.

Thus, Norbert's Contrast goes further to specify that all systems are balancing and coordinating multiple perspectives—some distant from the actual point of work and some local at points of contact with the monitored process (Figures 2 and 3). This means that JCSs are poly-centric, with multiple centers for control in parallel, which need to be coordinated in a dynamic balance (Ostrom, 1999; Woods & Shattuck, 2000; Ostrom, 2003; Woods, 2006b). A center of control has partial or bounded scope of authority for adapting to meet sub-goals within a context of other centers. "Local" centers make direct contact with sources of variability, which give them privileged ability to pick up on surprises, disruptions, and opportunities to plans in progress. "Distant" centers of control provide broader perspectives over time, space, and multiple functions, which allows them to see how to coordinate activities to achieve larger goals under tighter pressures. Control is poly-centric when overall performance results from interactions across activities in different areas and overall goals are achieved through work to achieve partial goals at each center.

The coordination between distant models and local action arises because there is differential responsibility over the multiple goals and because the potential for events to challenge the model of practice built into distant plans varies. The coordination in poly-centric systems is necessary to balance the tradeoff between risks of under- and over-adaptation when events arise that could challenge boundary conditions (military structures become poly-centric when faced with high tempo and adaptive adversaries; e.g., processes of commander's intent as in Shattuck & Woods, 2000).

The above set of generalizations represents an attempt to present first principles about work as processes of coping with complexity. These generalizations provide a check on the risk of over-simplifications when studying, modeling, and designing JCSs at work (the tendencies in Table 4). These "laws" represent a tentative functional synthesis that can be used to jump-start and guide studies of JCSs. The observations that result from these studies should lead to new findings-to-be-explained, new debates over possible explanations, and revised or new generalizations about work.

GENERIC REQUIREMENTS TO DESIGN JOINT COGNITIVE SYSTEMS THAT WORK

Cognitive Systems Engineering has discovered tentative first principles about JCSs at work. But the goal of CSE is to help develop JCSs that work, i.e., to contribute to envisioning promising directions for using the capabilities of technology and to avoid undesirable side effects. What then does past research in CSE tell us about how to support practitioners in context? This is a version of a classic dilemma across all aspects of inquiry about human factors—how is research valuable to the design and development of new systems?

Sometimes, researchers respond, the answer lies in methods for discovering the demands of a work setting and the strategies of experts. This diagnosis suggests that effort should be devoted to streamlining and transferring methods to a broader pool of people involved in processes of development and fielding of new artifacts. Other times, researchers respond that the answer is to translate empirical results into design guidance that participants from other disciplines can use when they are embedded in the processes of development and fielding of new artifacts.

But CSE goes beyond these approaches to utilize a third strategy. Results in CSE provide generic patterns that can seed and guide the search for what would be useful—supporting requirements discovery in design and development (Woods, 1998; Roesler et al., 2005). Ultimately, the test of CSE as a research strategy is its ability to identify basic requirements for how to support work that must be met if new technology will be useful to practitioners in context.

Based on first principles or laws about joint cognitive systems as captured tentatively in the section before, the field has developed a set of generic requirements or support functions. These generalizations are useful in design and evaluation in several ways:

• They apply across different specific settings, since they are domain independent or generic.
• They guide/test cognitive work analyses of specific settings, since they are general findings.
• They specify basic requirements to be met if artifacts-to-be-realized are to support JCSs at work.
• They seed design with initial hypotheses about what will prove useful.
• They focus development resources quickly on high leverage areas.
• They provide criteria for testing for success in evaluation studies.

The search for affordance in design is a kind of functional claim about the relationship of artifact to practitioner to demands, given goals (see pp. 63-64). As we have seen, claims of affordance are based on models of demands and strategies for coordination and resilience. To achieve high levels of performance in a JCS at work (in the sense of control—the ability to adapt behavior in changing circumstances to pursue a goal—the key for the first family of the Laws), new systems need to support coordination and resilience.

Support for Coordination—the ability to coordinate and synchronize activity across agents in pace with changing situations.

This includes issues about building common ground or a shared frame of reference (e.g. the updating case on p. 91-92; Olson et al., 2001); providing anticipatory cues and landmarks to synchronize assessments and activities (Nyssen & Javaux, 1996); providing the means for team work with machine agents given the fundamental collaborative asymmetry (Klein et al., 2004); supporting inter-predictability across distributed agents and sharing of intent in effective joint activity, developing a shared commitment to align and balance multiple goals (Klein, Feltovich et al., 2005).

Support for Resilience—the ability to anticipate and adapt to potential for surprise and error.

This includes issues about

• how practice includes, tests, and revises failure sensitive strategies;
• how broadening checks and cross checks are used to explore outside current boundaries, set, focus, or priority to avoid premature narrowing and overcome the brittleness of automata (Hollnagel et al., 2006).

One challenge is that of reframing—not getting stuck in one view or model. Another challenge is to find ways to balance pressure to narrow/simplify with comprehensiveness and broadening checks, given limited time and resources.

Table 6 presents the generic requirements that capture what it means to provide affordances for work in JCSs and illustrates some of the sub-functions for each of the main support requirements: Observability (feedback that provides insight into a process), Directability (ability to re-direct resources, activities, priorities as situations change and escalate), Control of Attention (ability to re-orient focus in a changing world), and Shifting Perspectives (coordinating contrasting points of view). Working to meet these generic requirements can be seen as collecting a set of findings on how to tame and manage new complexities and tighter demands so that control is amplified. Designing and testing JCSs to these requirements helps us to avoid being trapped into local coping mechanisms such as the oversimplification tendency to muddle through competing demands.

This set of generic functions has already proven quite useful in current work on how to make team players of automata. (Sarter, 2002; Klein et al., 2004) and in considering how to support information analysis (Patterson et al., 2001; Elm et al., 2005).

Table 6. Generic Requirements to Support JCSs that Work

Support for Observability: feedback that provides insight into a process
• Integrate data based on a model of the process
• Align data to reveal patterns and relationships in a process
• Provide context around details of interest
• Overcome "keyhole"/extend peripheral awareness
• See sequence & evolution over time
• See future activities & contingencies
• Decompose integrations and inferences into sources, process, base evidence

Support for Directability: ability to direct/re-direct resources, activities, priorities as situations change and escalate
• Anticipation/projection
• Models of capability
• Policies for adaptation
• Intent communication

Support for Directing Attention: ability to re-orient focus in a changing world
• Track others' focus of attention
• Judge interruptibility of others
• Use Pre-attentive reference

Support for Shifting Perspectives: contrasting points of view
• Seeding—structure & kick start initial activity
• Reminding—suggest other possibilities as activity progresses
• Critiquing—point out alternatives as activities come to a close

The four generic requirements have been evident throughout the stories running through this book. Observability has proven to be the basic first step for support of work (pp. 83-84), and one of the first themes in CSE was how to provide better feedback to see the emerging effects of decisions, actions, and policies (Goodstein, 1981; Woods, 1984; Rasmussen, 1985; see pp. 136-137). In general, demands that increase complexity can be balanced or tamed through improved observability (Billings, 1987; Christoffersen & Woods, 2002). To design feedback that improves observability requires innovation so that external representations are (Woods, 1995; Woods et al., 2002; Potter et al., 2003):

- Integrated to capture relationships and to make patterns emerge, not simply to provide access to a large set of base data or elements
- Event based to capture change and sequence, not simply to capture the current values on each data channel
- Future oriented to help people assess what could happen next, not simply what has happened
- Context-sensitive and tuned to the interests and expectations of the monitor.

Note that designing for observability requires finding models that serve as the base for integrating data over functions, space and time, though this is subject to the "right?" model catch of family 2 of the laws and the representation effect (Norman, 1993).

Once observability is established, design can shift focus to consider the other basic support functions. Directability is required as a consequence of the laws of coordination to avoid the trap of "either you do it all or I do it all" systems (see the stories on pages 113-116). The Global Hawk mishap story was one story illustrating the costs of low directability. Design then needs to provide some capacity for modifying and re-directing the actions and targets of the other parties in a joint activity as conditions and priorities change. Progress on developing means for directability has lagged due to the corrosive impact of the substitution myth, and to the erroneous assumption that the next levels of autonomy to be reached will eliminate the need for coordination. Promising directions have been identified based on regulating policies that govern agent behavior (Bradshaw et al., 2004) or based on intent communication between distant supervisors and local actors (Shattuck & Woods, 2000).

Directed attention or control of attention is concerned with the problem of how to help decision makers re-orient attention to new significant information even when they are busy with other tasks. It is critical for skills of anticipation (pp. 80-85). This generic support function appeared in the very first story of JCSs at work (pp. 15-16). It addresses the criteria of how to direct the attention of human monitors to "interesting" behaviors or changes in behavior. It also addresses the criteria of how to cull out unimportant data and avoid the debilitating consequences of false alerts. The design of artifacts such as alerts and alarms can either hobble or support effective control of attention in any multi-threaded situation (Sarter, 2002). Sarter's demonstrations of how to use tactile cues to solve the problem of mode awareness on flight decks is one illustration of how to create affordances for directed attention (Sklar & Sarter, 1999; Nikolic & Sarter, 2001).

Interestingly, people are fundamentally able to assess where another person's attention is focused in order to coordinate when and how to inform them about new events (Moore & Dunham, 1995), e.g., judging interruptibility is a basic aspect of collaboration. To achieve coordination, control of attention requires bi-directional re-focusing, in which one agent can influence where and on what the human monitor focuses, and where the human monitor can choose to focus and still receive support from other (human or machine) agents to recognize ongoing events at different scales.

Shifting perspectives has turned out to be a fundamental aspect of JCSs. The basic role of different perspectives was captured by Bartlett (1932, p. 4):

We may consider the old and familiar illustration of the landscape artist, the naturalist, and the geologist who walk in the country together. The one is said to notice and recall beauty of scenery, the other details of flora and fauna, and the third the formations of soils and rocks. In this case, no doubt, the stimuli being selected in each instance from what is present, are different for each observer, and obviously the records made in recall are different also. Nevertheless, the different reactions have a uniformity of determination, and in each case spring from established interests.

Aspects of perspective run throughout the patterns on JCSs at work in this book. Revising assessments, or avoiding getting stuck in one view (fixation), is all about supporting shifts in perspective. Cross checks for resilience are about providing means to coordinate and contrast different perspectives provided by agents in different roles (e.g., Patterson et al., 2004). Visual momentum design techniques are about helping people move between different perspectives on data, and serve as a reminder to design the different perspectives into the system for display of data (Woods & Watts, 1997). Processes of building common ground are, at heart, a process for monitoring and exchanging perspectives (and therefore show how common ground is not about providing all agents with the same view) as failed to occur in the replanning breakdown described on pp. 139-140. Seeing where someone else is gazing to ascertain where their attention is focused is a fundamental property of cognition that is tuned to being able to assess another's perspective. Similarly, judging interruptibility is about assessing another's perspective (where perspective is more abstract). Stance in cooperative work is about another group's perspective—perspective in an even more abstract sense. Crossing perspectives is commonplace in distributed work as people combine some core of similar understanding with their unique perspectives due to their own backgrounds, expertise, and point of view on the work. All of this adds flexibility, cross checks, and novel approaches (Hong & Page, 2002). CSE itself arose as a new perspective on work that stepped outside of and integrated component perspectives on people, on technology and on specific work settings (as captured in the story of envisioning pp. 144-149).

These four generic requirements for support (affordances) are orchestrated to achieve coordination and resilience in JCSs. Coordination across agents begins with the need to establish "common ground" or a shared frame of reference about the events in the world. This is a form of developing support for observability. One can then develop criteria and test for what forms of observability support building common ground. For example, several lines of recent work emphasize the need to capture temporal patterns (see p. 80-85; Zacks, 2004; Christoffersen et al., 2006).

Coordination across agents (and across human and machine agents) also is built upon shifting perspectives as different people in different roles have different views, knowledge and stances about the situation in the world. For example, in effective human-computer agent coordination in information analysis, the results from different computer algorithms define different perspectives across which the human monitor can shift or contrast (Woods & Watts, 1997).

Shifting perspectives includes roles for agents to help broaden another monitor's view of the situation and trends (Smith et al., 2001; 2004). This has proved critical in avoiding or escaping fixations (pp. 76-77). Resilience is enhanced though collaborative broadening that juxtaposes multiple perspectives in ways that make apparent what is hidden or downplayed in any single perspective. Juxtaposition of perspectives can occur through different combinations of *seeding*—helping to structure & kick start initial activity; *reminding*—suggesting other possibilities as analysis progresses; *critiquing*—point out alternatives as analysis matures to a conclusion (Layton et al., 1994; Guerlain et al., 1999; Smith et al., 1997). Seeding, reminding, and critiquing are basic *collaborative broadening* functions that help us explore outside current boundaries, set, focus, or priority to avoid premature narrowing (Patterson et al., 2001).

In the case of *Team work with machine agents,* research shows that support is required for observability functions such as common ground—shared frame of reference to make other agents' models, intent, and activities observable. This applies to event observability—what events appear to be ongoing in the world. Event observability tests the criteria of (a) how to model complex events against a changing background of activity and at different time scales, (b) how to integrate inputs from data sources that vary in modality and in level of autonomy, and (c) how to combine multiple information sources into a coherent picture for human monitors. It also applies to algorithm observability, which makes machine agents' models, intent, and activities observable to human problem-holders—e.g., what is detectable with what boundary conditions by surveillance algorithms.

Research on *Team work with machine agents* also found that effective JCSs require support for directability functions—how to re-direct agent resources as situations change (delegation). Algorithm directability provides the means for human monitors or problem-holders to re-direct and interact with partially autonomous systems. The greater the autonomy of automated systems, the higher the need for responsible people to be able to re-direct the automata as high level resources. For example, as the machine decides how to reconfigure its sensor resources given its targets for "coverage," or as it decides to gather more "information" for the human monitor, the human monitor will need to redirect the reconfiguration process when situations arise outside the machine's knowledge (brittleness), and where the human has information that is not available to the algorithms.

Design Responsibility

At this stage, it is important to go back to previous discussions of how design plays a role in CSE. First, the discussion of core values in CSE reminded us that researchers also have responsibility as participants in creating future work systems with other stakeholders (the fourth value on pp. 6-7). Second, the discussion of studies to discover how JCSs work ended with challenges created by the necessity for these studies to feed design processes—in particular, the search for what would be useful given the processes of adaptation and change at work (pp. 57-58). Third, the patterns in coordination and resilience demonstrate that new technology stimulates change in JCSs and triggers a complex set of reverberations. Fourth, new technology also provides power, which is necessary but not sufficient to amplify control in designing future JCSs that work. Fifth, a theme running through the patterns is that people provide a unique resource in JCSs given that they are meaning-seeking, learning, context-sensitive, coordinating (social), and responsible agents. Sixth, understanding JCSs that work provides insight about how to cope with the complexities, tradeoffs and dilemmas that arise from the need to act in changing, finite resource, uncertain, conflicted worlds of work.

These points lead us to a basic principle of design responsibility (Woods, 2002):

There is no neutral in design:
In design, we either hobble or support people's natural ability to express forms of expertise.

People can learn to express various forms of expertise with experience, and they will shape artifacts and interactions to assist them as they cope with demands of work to meet goals. Thus their work will be "well" adapted given the constraints and degrees of freedom for action within their roles. In re-designing such adapted systems, we provide opportunities for new forms of adaptation to better fit the changing demands of practice and on practice (Winograd & Flores, 1986). Design is an intervention into these ongoing worlds of activity (Flores et al., 1988); the adaptive responses within fields of practice provide feedback about the hypotheses of what would prove useful, embodied in the form of new artifacts.

As the stories of intervention and adaptation to amplify control illustrate, research results participate in design. Research inputs have real effects on those at the sharp end, squeezing or enhancing their ability to express expertise. If the research is meaningful, it participates in these processes. Thus, researchers have to accept their design responsibility—there is no neutral point to stop at or neutral screen to duck behind in the search for affordances to support work.

The new technologies for autonomy, connectivity, data collection/ transmission/availability provide designers with power and with freedom. Design responsibility means that research also has to provide results that specify how to use that power to support, not hobble, the expression of expertise. The tentative laws that govern JCSs at work and the generic requirements to design JCSs that work provide the basis for control of the powers of new technology and to harness them for innovation in pursuit of the question—*what will be useful?*

Design, then, is a form of work best seen (though at another scale) as a control process, often quite scrambled. Feedback on the impact of introducing designed objects in ongoing fields of practice stimulate processes of adaptation and control in the face of variability and potential for surprise.

PATTERNS AND STORIES

This book provides stories of JCSs at work. The set is not complete as the expanding research base discovers more patterns about resilience, coordination, and affordance. New work challenges old generalizations, and sets in motion new functional syntheses. As a pattern book in Alexander's sense (Alexander et al., 1977), this book is intended to stimulate the process of finding, sharing, debating, revising, and discovering general results that emerge at the intersection of people, technology, and work.

Approaching research results as Alexander patterns (pp. 11-12) challenges conventional notions about how to represent findings. Patterns about work are storylines described in terms of intervention, adaptation and transformation. Representing research as a collection of such story archetypes is a radical departure from traditions. But notice how each generalization phrase or label is an encapsulation of storylines packed up and ready to burst out again. *Coping with complexity* implies a story triggered by some challenging demand, and anticipates how a protagonist meets or works around the challenge. *Over-simplification* sets us up to expect a trap as a usual shortcut hinders the protagonist's ability to handle a different situation. Only the audience for the story has the necessary broader perspective to see the need to revise and re-frame. Just the label *automation surprise* suggests a story of communication breakdown where the machine agents suggest they have understood the human's instructions, yet what it will do does not match what the human intended; again, the audience, but not the participants, has the broader perspective to see opportunities for repair slide by as the vehicle behaves other than expected and heads closer to danger. These storylines play out in great variety in specific settings and with each different wave of technology, yet we can still recognize the general storyline underway.

The storylines occur in parallel at multiple levels. We can re-use the storylines to help us understand a specific sharp end of practice as well as to characterize how design gets trapped into developing clumsy automation. We can approach new settings listening carefully to hear which story archetypes are present. Recognizing the story lines there helps us anticipate where that work setting is heading, as once a story is recognized as underway, one can then project the relevant trajectories encapsulated in the archetype.

Even Neisser's perceptual cycle is a story archetype (*JCS-Foundations*, p. 20 and Figure 1.7). CSE does not simply start with Neisser's perceptual cycle as a conceptual framework to abstract stories of JCSs at work. CSE as an approach to re-engineer work is itself engaged in Neisser's perceptual cycle. CSE uses the JCS concept as a frame or perspective on diverse work activities and settings. This frame helps one to notice phenomena that emerge at the intersection of people, technology and work in terms of recurring patterns in resilience, coordination, and affordance. By connecting to these frames, some otherwise diverse details and events cohere as variations around the general pattern, while other details and events stand out since they depart from expectations as anomalies that require revision, exploration, and explanation. These patterns generate promising directions for intervention as tentative hypotheses. Observing the impact of such interventions provides feedback to revise the framing concepts that had generated the hypothesis.

The complementarity between the universal and the particular fundamental to Neisser's perceptual cycle is the engine for CSE. Moving from abstractions about work in general to make authentic contact with particular scenarios, practitioners and artifacts, and then stepping through the particulars in that setting to see the operation of general patterns and story archetypes is the creative tension that drives understanding how JCSs work and innovating JCSs that work.

To summarize: Work in CSE has abstracted basic patterns or regularities that recur in many specific settings but that transcend the details of that setting. These general patterns help focus new studies of JCSs, help jump start design, and help avoid repeating past design failures.

Additional resources are the stories of JCSs at work generated and to be generated that capture patterns in coordination, resilience and affordance. From this base we will continue to develop new stories that discover promising directions for innovating future JCSs that work.

Bibliography

Abbott, K. H. (1990). *Robust fault diagnosis of physical systems in operation.* Unpublished Doctoral dissertation, State University of New Jersey, Rutgers.

Abbott, K. H. (1996). The interfaces between flightcrews and modern flight deck systems. Washington D. C.: Federal Aviation Administration, June 18, 1996.

Ackoff, R. L. (1979). The future of operational research is past. *Journal of the Operational Research Society, 30*, 93–104.

Alexander, C., Ishikawa, S. & Silverstein, M. (1977). *A pattern language.* New York: Oxford University Press.

Ash, J. S., Berg, M. & Coiera, E. (2004). Some unintended consequences of information technology in health care: The nature of patient care information system-related errors. *Journal of the American Medical Informatics Association, 11*, 104-112.

Ashby, W. R. (1957). *An introduction to cybernetics.* London: Methuen.

Barley, S. & J. Orr. J., (Eds.), (1997). *Between craft and science: Technical work in US settings.* Ithaca, NY: IRL Press.

Bartlett F. C. (1932). *Remembering: A study in experimental and social psychology.* Cambridge: Cambridge University Press.

Billings, C. E. (1997). *Aviation automation: The search for a human-centered approach.* Hillsdale, NJ: Lawrence Erlbaum.

Bourrier, M. (1999). Constructing organizational reliability: the problem of embeddedness and duality. In J. Misumi, B. Wilpert & R. Miller (Eds.), *Nuclear safety: A human factors perspective.* Philadelphia: Taylor & Francis, 1999.

Bradshaw, J. M. (Ed.). (1997). *Software agents.* Cambridge, MA: AAAI Press/The MIT Press.

Bradshaw, J. M., Feltovich, P., Jung, H., Kulkarni, S., Taysom, W., & Uszok, A. (2004). Dimensions of adjustable autonomy and mixed-initiative interaction. In M. Nickles, M. Rovatsos, & G. Weiss (Eds.), *Agents and computational autonomy: Potential, risks, and solutions.* Lecture Notes in Computer Science, Vol. 2969. Berlin, Germany: Springer-Verlag.

Brennan, S. E. (1998). The grounding problem in conversations with and through computers. In S. R. Fussel & R. J. Kreuz (Eds.), *Social and cognitive psychological approaches to interpersonal communication.* Hillsdale, NJ: Lawrence Erlbaum.

Brown, J. P. (2005a). Ethical dilemmas in health care. In M. Patankar, J. P. Brown & M. D. Treadwell (Eds.), *Safety ethics: Cases from aviation, health care, and occupational and environmental health.* Burlington VT: Ashgate.

Brown, J. P. (2005b). Key themes in health care safety dilemmas. In M. Patankar, J. P. Brown & M. D. Treadwell (Eds.), *Safety ethics: Cases from aviation, health care, and occupational and environmental health.* Burlington VT: Ashgate.

Bruner, J. (1986). *Actual minds, possible worlds.* Cambridge, MA: Harvard University Press.

Bruner, J. (1990). *Acts of meaning.* Cambridge, MA: Harvard University Press.

Burke, J. L., Murphy, R. R., Coovert, M. D., & Riddle, D. L. (2004). Moonlight in Miami: A field study of human robot interaction in the context of an urban search and rescue disaster response training exercise. *Human Computer Interaction, 19*, 85-116.

CAIB (*Columbia* Accident Investigation Board). (2003). Report, 6 vols. Government Printing Office, Washington, DC. www.caib.us/news/report/default.html.

Carroll, J. M. (1997). Scenario-based design. In M.G. Helander, T.K. Landauer, & P. Prabhu (Eds.). *Handbook of human-computer interaction* (2nd edition), Amsterdam: Elsevier Science.

Carroll, J. M., Kellogg, W. A. & Rosson, M. B. (1991). The task-artifact cycle. In J. M. Carroll (Ed.), *Designing interaction: Psychology at the human-computer interface*, Cambridge: Cambridge University Press.

Carroll, J. M., Neale, D. C., Isenhour, P. L., Rosson, M. B., & McCrickard, D. S. (2003). Notification and awareness: synchronizing task-oriented collaborative activity. *International Journal of Human-Computer Studies, 58*, 605-632.

Casper, J. & Murphy, R. R. (2003). Human-robot interaction during the robot assisted urban search and rescue response at the World Trade Center. *IEEE SMC B, 33*, 367-385.

Cassirer, E. (1953). *The philosophy of symbolic forms, Vol. 1: Language.* Yale University Press, New Haven CT. (Translated by R. Manheim, original work published 1923).

Cawsey, A. (1992). *Explanation and interaction.* Cambridge, MA: MIT Press.

Chandrasekaran, B., Tanner, M. C. & Josephson, J. (1989). Explaining control strategies in problem solving. *IEEE Expert, 4*, 9-24.

Chinn, C.A. & Brewer, W.F. (1993). The role of anomalous data in knowledge acquisition: A theoretical framework and implications for science instruction. *Review of Educational Research, 63*, 1-49.

Chow, R., Christoffersen, K. & Woods, D.D. (2000). A Model of Communication in Support of Distributed Anomaly Response and Replanning. In Proceedings of the IEA 2000/HFES 2000 Congress, Human Factors and Ergonomics Society, July, 2000.

Christoffersen, K. & Woods, D. D. (2002). How to make automated systems team players. In E. Salas (Ed.), *Advances in Human Performance and Cognitive Engineering Research, Volume 2.* St. Louis, MO: Elsevier Science.

Christoffersen, K. & Woods, D. D. (2003). Making Sense of Change: Extracting Events From Dynamic Process Data. Institute for Ergonomics/Cognitive Systems Engineering Laboratory Report, ERGO-CSEL 01-TR-02. September 25, 2003.

Christoffersen, K., Woods, D. D. & Blike, G. (2006). Discovering the events expert practitioners extract from dynamic data streams: The mUMP Technique. Submitted.

Clancey, W. J. (1983). The epistemology of a rule-based expert system: A framework for explanation. *Artificial Intelligence, 20*, 215-251.

Clancey, W. J. (1997). *Situated cognition: On human knowledge and computer representations.* Cambridge, MA: Cambridge University Press.

Clancey, W. J. (2006). Participant observation of a mars surface habitat mission simulation. *Habitation*, in press.

Clark, H. H. & Brennan, S. E. (1991). Grounding in communication. In L. B. Resnick, J. M. Levine & S. D. Teasley (Eds.), *Perspectives on socially shared cognition*. Washington D.C.: American Psychological Association.

Conant, R. C. (1976). Laws of information which govern systems. *IEEE Transactions on Systems, Man, and Cybernetics*, SMC-6, 240-255.

Conant, R. C. & Ashby, W. R. (1970). Every good regulator of a system must be a model of that system. *International Journal of Systems Science*, *1*, 89-97.

Cook, R.I., Woods, D.D. & McDonald, J.S. (1991). Human Performance in Anesthesia: A Corpus of Cases. Cognitive Systems Engineering Laboratory Report, prepared for Anesthesia Patient Safety Foundation, April 1991.

Cook, R.I., Potter, S., Woods, D.D. & McDonald, J.S. (1991). Evaluating the human engineering of microprocessor controlled operating room devices. *Journal of Clinical Monitoring*, *7*, 217-226.

Cook, R. I., Woods, D.D. & Howie, M.B. (1992). Unintentional delivery of vasoactive drugs with an electromechanical infusion device. *Journal of Cardiothoracic and Vascular Anesthesia*, *6*, 238-244.

Cook, R. I. & Woods, D.D. (1996). Adapting to new technology in the operating room. *Human Factors*, *38*, 593-613.

Cook, R. I. (1998). Being Bumpable. Presentation at Fourth Conference on Naturalistic Decision Making. Warrenton, VA, May 29-31, 1998.

Cook, R.I., Woods, D.D. & Miller, C. (1998). *A Tale of Two Stories: Contrasting Views on Patient Safety*. Chicago IL: National Patient Safety Foundation, April 1998 (available at www.npsf.org/exec/report.html).

Cook, R.I., Render, M. L. & Woods, D.D. (2000). Gaps in the continuity of care and progress on patient safety. *British Medical Journal*, *320*, 791—794.

Cook, R. I. & O'Connor, M. (2005). Thinking about accidents and systems. In H. Manasse & K. Thompson (Eds.), *Medication safety: A guide to health care facilities*. Bethesda, MD: American Society of Health-System Pharmacists.

Cordesman, A. H. & Wagner, A. R. (1996). *The lessons of modern war, Vol.4: The Gulf War*. Boulder, CO: Westview Press.

Craik, K. J. W. (1947). Theory of the operator in control systems: I. The operator as an engineering system. *British Journal of Psychology*, *38*, 56-61.

De Keyser, V. (1990). Temporal decision making in complex environments. *Philosophical Transactions of the Royal Society of London, B 327*, 569-576.

De Keyser, V. & Woods, D.D. (1990). Fixation errors: Failures to revise situation assessment in dynamic and risky systems. In A.G. Colombo & A. Saiz de Bustamante, (Eds.), *Systems reliability assessment*, Dordrechts, The Netherlands: Kluwer Academic.

De Keyser, V. (1992). Why field studies? In M. G. Helander & N. Nagamachi (Eds.), *Design for manufacturability: A systems approach to concurrent engineering and ergonomics*. London: Taylor & Francis.

De Keyser, V. & Samercay, R. (1998). Activity theory, situated action and simulators. *Le Travail Humain*, *61*, 305-312.

Dekker, S. W. A. (2002). The field guide to human error investigations. London: Ashgate.

Dekker, S. W. A. (2004). Ten questions about human error: A new view of human factors and system safety. Hillsdale, NJ: Lawrence Erlbaum.

Dominguez, C., Flach, J., McDermott, P., McKellar, D., & Dunn, M. (2004). The conversion decision in laparoscopic surgery: Knowing your limits and limiting your risks. In K. Smith, J. Shanteau & P. Johnson (Eds.), *Psychological investigations of competence in decision making.* New York: Cambridge University Press.

Dörner, D. (1983). Heuristics and cognition in complex systems. In R. Groner, M. Groner & W.F. Bischof (Eds.), *Methods of heuristics*, Hillsdale, NJ: Lawrence Erlbaum.

Doyle, J. D. & Vicente, K. J. (2001). Patient-controlled analgesia. *Canadian Medical Association Journal, 164,* 620.

Duchon, A. P. & Warren, W. H. (2002). A visual equalization strategy for locomotor control: Of honeybees, robots and humans. *Psychological Science, 13,* 272-278.

Dugdale, J., Pavard, B. & Soubie, J. L. (2000). A pragmatic development of a computer simulation of an emergency call centre. In G. De Michelis et al. (Eds.), *Designing cooperative systems*, Proceedings of COOP'2000. Amsterdam: IOS Press.

Elm, W., Potter, S., Tittle, J., Woods, D., Patterson, E., & Grossman, J. (2005). Finding decision support requirements for effective intelligence analysis tools. In Proceedings of the Human Factors and Ergonomics Society 49th Annual Meeting. 26-28 September, Orlando FL.

Feltovich, P. J., Spiro, R. J., & Coulson, R. L. (1997). Issues of expert flexibility in contexts characterized by complexity and change. In P. J. Feltovich, K. M. Ford, & R. R. Hoffman (Eds.), *Expertise in context: Human and machine.* Menlo Park, CA. AAAI/MIT Press.

Feltovich, P. J., Coulson, R. L., & Spiro, R. J. (2001). Learners' (mis)understanding of important and difficult concepts: A challenge to smart machines in education. In K. D. Forbus & P. J. Feltovich (Eds.), *Smart machines in education.* Menlo Park, CA: AAAI/MIT Press.

Feltovich, P. J., Hoffman, R. R., Woods, D., & Roesler, A. (2004). Keeping it too simple: How the reductive tendency affects cognitive engineering. *IEEE Intelligent Systems, 19,* 90-94.

Feltovich, P. J., Bradshaw, J. M., Jeffers, R., Suri, N., & Uszok, A. (2004). Social order and adaptability in animal and human cultures as analogues for agent communities: Toward a policy-based approach. In A. Omacini, P. Petta & J. Pitt (Eds.), *Engineering societies in the agents world IV* (Lecture Notes in Computer Science Series). Berlin: Springer-Verlag.

Fetterman, D. M. (1989). *Ethnography: step by step.* Beverly Hills, CA: Sage Publications.

Fischer, U., & Orasanu, J. (2000). Error-challenging strategies: Their role in preventing and correcting errors. In Proceedings of the International Ergonomics

Association 14th Triennial Congress and Human Factors and Ergonomics Society 44th Annual Meeting in San Diego, California, August 2000.

Fitts, P. M., Ed. (1951). Human Engineering for an Effective Air Navigation and Traffic Control System. National Research Council, Washington (also Columbus, OH: Ohio State University Research Foundation).

Flach, J. M., Jacques, P. F., Patrick, D. L., Amelink, M., Van Paassen, M. M. & Mulder, M. (2003). A search for meaning: A case study of the approach-to-landing. In E. Hollnagel (Ed.), *Handbook of cognitive task design*. Hillsdale, NJ: Lawrence Erlbaum.

Flach, J. M., Smith, M. R. H., Stanard, T., & Dittman, S. M. (2004). Collision: Getting them under control. In H. Hecht & G.J.P. Savelsbergh (Eds.) *Theories of time to contact*. Amsterdam: Elsevier.

Flach, J.M., Dekker, S. W. A. & Stappers, P. J. (2006). Playing twenty questions with nature: Reflections on quantum mechanics and cognitive systems. *Theoretical Issues in Ergonomic Science*, submitted.

Flores, F., Graves, M., Hartfield, B. & Winograd, T. (1988). Computer systems and the design of organizational interaction. *ACM Transactions on Office Information Systems, 6*, 153-172.

Folk, C. L., Remington, R. W. & Johnston, J. C. (1992). Involuntary covert orienting is contingent on attentional control settings. *Journal of Experimental Psychology, Human Perception and Performance, 18*, 1030-1044.

Freed, M., Bonasso, P., Ingham, M., Kortenkamp, D., Pell, B. & Penix, J. (2004). Trusted autonomy for spaceflight systems. Proceedings of American Institute of Aeronautics and Astronautics.

Gaba, D. M., Maxwell, M. & DeAnda, A. (1987). Anesthetic mishaps: Breaking the chain of accident evolution. *Anesthesiology, 66*, 670-676.

Galison, P. L. (1997). *Image and Logic*. Chicago: University of Chicago Press.

Garrett, S. K., & Caldwell, B. S. (2002). Mission Control Knowledge Synchronization: Operations To Reference Performance Cycles. Proceedings of the Human Factors and Ergonomics Society 46th Annual Meeting, Baltimore, MD.

Gaver, W. W. (1997). Auditory interfaces. In M. Helander, T. K. Landauer, & P. Prabhu (Eds.), *Handbook of human-computer interaction* (2nd edition), Amsterdam: Elsevier Science.

Getty D. J., Swets, J. A., Pickett, R. M., & Gonthier, D. (1994). System operator response to warnings of danger. *Journal of Experimental Psychology: Applied 1, 19-33*.

Gettys, C. F. & Fisher, S. D. (1979). Hypothesis plausibility and hypothesis generation. *Organizational Behavior and Human Performance, 24*, 93-110.

Gettys, C. F., Mehle, T. & Fisher, S. D. (1986). Plausibility assessments in hypothesis generation. *Organizational Behavior and Human Decision Processes, 37*, 14-33.

Gettys, C. F., Pliske, R. M., Manning, C. & Casey, J. T. (1987). An evaluation of human act generation performance. *Organizational Behavior and Human Decision Processes, 39*, 23-51.

Gibson, J. J. (1979). *An ecological approach to perception.* Boston: Houghton Mifflin.

Ginsburg, G. P. & Smith, D. L. (1993). Exploration of the detectable structure of social episodes: The parsing of interaction specimens. *Ecological Psychology,* 5(3), 195-233.

Goodstein, L. (1981). Discriminative display support for process operators. In J. Rasmussen & W. Rouse (Eds.), *Human detection and diagnosis of system failures.* New York: Plenum Press.

Gopher, D. (1992). The skill of attention control: Acquisition and execution of attention strategies. In D. E. Meyer & S. Kornblum, (Eds.), *Attention and performance XIV: Synergies in Experimental Psychology, Artificial Intelligence and Cognitive Neuroscience,* Cambridge MA: MIT Press.

Guerlain, S., Smith, P.J., Obradovich, J.H., Rudmann, S., Strohm, P., Smith, J., & Svirbely, J. (1996). Dealing with brittleness in the design of expert systems for immunohematology. *Immunohematology, 12,* 101-107.

Guerlain, S., Smith, P.J., Obradovich, J. H., Rudmann, S., Strohm, P. Smith, J.W., Svirbely, J., & Sachs, L. (1999). Interactive Critiquing as a Form of Decision Support: An Empirical Evaluation. *Human Factors, 41,* 72-89.

Heath, C. & Luff, P. (1992). Collaboration and control: Crisis management and multimedia technology in London Underground line control rooms. *Computer-Supported Cognitive Work, 1,* 69-94.

Heider, F. & Simmel, M. A. (1944). An experimental study of apparent behavior. *American Journal of Psychology 57,* 243-59.

Hirschhorn, L. (1993). Hierarchy vs. bureaucracy: The case of a nuclear reactor. In K. H. Roberts (Ed.), *New challenges to understanding organizations.* New York: McMillan.

Hirschhorn, L. (1997). Quoted in Cook, R. I., Woods, D. D. & Miller, C. (1998). *A tale of two stories: Contrasting views on patient safety.* National Patient Safety Foundation, Chicago IL, April 1998 (available at www.npsf.org).

Ho, C-Y., Nikolic. M., Waters, M., & Sarter, N. B. (2004). Not now: Supporting interruption management by indicating the modality and urgency of pending tasks. *Human Factors, 46,* 399-409.

Hochberg, J. (1986). Representation of motion and space in video and cinematic displays. In K. R. Boff, L. Kaufman, & J. P. Thomas, (Eds.), *Handbook of human perception and performance, Vol. I.* New York: John Wiley & Sons.

Hoffman, R. R., Crandall, B. W., & Shadbolt, N. R. (1998). Use of the critical decision method to elicit expert knowledge: A case study in cognitive task analysis methodology. *Human Factors, 40,* 254-276.

Hoffman, R. & Woods, D.D. Studying cognitive systems in context. *Human Factors, 42,* 1-7, 2000.

Hoffman, R.R., Ford, K. M., Feltovich, P.J., Woods, D.D., Klein, G. & Feltovich, A. (2002). A rose by any other name ... would probably be given an acronym. *IEEE Intelligent Systems,* July/August, 72-79.

Hogarth, R.M. (1986). Generalization in decision research, The role of formal models. *IEEE Systems, Man, and Cybernetics, SMC-16,* 439-449.

Hollnagel, E., Pedersen, O. & Rasmussen, J. (1981). Notes on Human Performance Analysis. Roskilde, Denmark: Risø National Laboratory, Electronics Department.

Hollnagel, E. & Woods, D. D. (1983). Cognitive Systems Engineering: New wine in new bottles. *International Journal of Man-Machine Studies*, 18, 583-600.

Hollnagel, E. (1993). *Human reliability analysis: Context and control.* London: Academic Press.

Hollnagel, E. (1998). *Cognitive reliability and error analysis method–CREAM.* Oxford: Elsevier Science.

Hollnagel, E. (1999). From function allocation to function congruence. In S. Dekker & E. Hollnagel (Eds.), *Coping with computers in the cockpit.* Aldershot, U.K.: Ashgate.

Hollnagel, E. (2001). Extended cognition and the future of ergonomics. *Theoretical issues in Ergonomics Science, 2*, 309-315.

Hollnagel, E. (2002). Time and time again. *Theoretical Issues in Ergonomics Science, 3*, 143-158.

Hollnagel, E. & Woods, D.D. (2005). *Joint Cognitive Systems: Foundations of Cognitive Systems Engineering.* Taylor & Francis.

Hollnagel, E., Woods, D.D. & Leveson, N. (2006). *Resilience Engineering: Concepts and precepts.* Aldershot, UK: Ashgate.

Hollnagel, E. & Woods, D.D. (2006). Epilogue: Resilience Engineering precepts. In Hollnagel, E., Woods, D.D. & Leveson, N., (Eds.), *Resilience Engineering: Concepts and precepts.* London: Ashgate.

Hong, L. & Page, S. E. (2002). Groups of diverse problem solvers can outperform groups of high-ability problem solvers. *Proceedings of the National Academy of Science: Economic Sciences.*

Hutchins, E. (1995a). *Cognition in the wild.* Cambridge, MA: MIT Press.

Hutchins, E. (1995b). How a cockpit remembers its speeds. *Cognitive Science, 19*, 265-288.

James, W. (1890). *Principles of psychology.* New York: H. Holt & Company.

Johannesen, L., Cook, R. I. & Woods, D.D. (1994). Cooperative communications in dynamic fault management. In Proceedings of the 38th Annual Meeting of the Human Factors and Ergonomics Society, October, Nashville, TN, 1994.

Johnson, P.E. & Thompson, W.B. (1981). Strolling down the garden path: Detection and recovery from error in expert problem solving. In Proceedings of the Seventh International Joint Conference on Artificial Intelligence, Vancouver, British Columbia.

Johnson, P.E., Moen, J.B. & Thompson, W.B. (1988). Garden path errors in diagnostic reasoning. In L. Bolec & M. J. Coombs (Eds.), *Expert System Applications.* New York: Springer-Verlag.

Johnson, P. E., Jamal, K. & Berryman, R. G. (1991). Effects of framing on auditor decisions. *Organizational Behavior and Human Decision Processes, 50*, 75-105.

Johnson, P. E., Grazioloi, S., Jamal, K. & Zualkernan, I. A. (1992). Success and Failure in Expert Reasoning. *Organizational Behavior and Human Decision Processes, 53*, 173-203.

Johnson, P. E., Grazioloi, S., Jamal, K. & Berryman, R. G. (2001). Detecting deception: Adversarial problem solving in a low base-rate world. *Cognitive Science, 25,* 355-392.

Josephson, J. & Josephson, S. (1994). *Abductive inference.* New York: Cambridge University Press.

Kelly-Bootle, S. (1995). *The Computer Contradictionary* (2nd Ed), Cambridge MA: MIT Press.

Kemeny, J.G. (1979). Report of the President's Commission on the accident at Three Mile Island. New York: Pergamon Press.

Kemp, R. (2005). Interview with Stan Correy, "What's the Data?" Australian Broadcast Corporation Radio National, August 28, 2005, http://www.abc.net.au/rn/talks/bbing/stories/s1445120.htm

Klein, G. A., Calderwood, R., & MacGregor, D. (1989). Critical decision method for eliciting knowledge. *IEEE Transactions on Systems, Man, and Cybernetics, 19,* 462-472.

Klein, G. A., & Crandall, B. W. (1995). The role of mental simulation in naturalistic decision making. In P. Hancock, J. Flach, J. Caird & K. Vicente (Eds.), *Local applications of the ecological approach to human-machine systems* (Vol. 2). Hillsdale, NJ: Lawrence Erlbaum.

Klein, G. (1999). *Sources of Power.* Cambridge, MA: MIT Press.

Klein, G., Ross, K., Moon, B., Klein, D., Hoffman, R. & Hollnagel, E. (2003). Macrocogniton. *IEEE Intelligent Systems,* May/June, 81-95

Klein, G., Woods. D.D., Bradshaw, J., Hoffman, R.R., & Feltovich, P.J., (2004). Ten Challenges for Making Automation a "Team Player" in Joint Human-Agent Activity. *IEEE Intelligent Systems,* November/December, 91-95.

Klein, G., Feltovich, P., Bradshaw, J. M. & Woods, D. D. (2005). Common ground and coordination in joint activity. In W. Rouse & K. Boff (Eds.). *Organizational simulation,* New York: Wiley.

Klein, G., Pliske, R., Crandall, B. & Woods, D. (2005). Problem Detection. *Cognition, Technology, and Work. 7,* 14-28.

Klein, G., Phillips, J. K., Rall, E., & Peluso, D. A. (in press). A data/frame theory of sensemaking. In R. R. Hoffman (Ed.), *Proceedings of the 6th International Conference on Naturalistic Decision Making.* Hillsdale, NJ: Erlbaum.

Koppel, R., Metlay, J., Cohen, A., Abaluck, B., Localio, A. R., Kimmel, S. & Strom, B. (2005). Role of computerized physician order entry systems in facilitating medication errors. *JAMA, 293(10),* 1197-1203.

Lanir, J. (1995). Agents of alienation. *Interactions,* July, 66-72.

Law, J. & Callon, M. (1995). Engineering and sociology in a military aircraft project: A network analysis of technological change. In S. L. Star, editor, *Ecologies of knowledge: Work and politics in science and technology.* Albany: State University of New York Press.

Layton, C., Smith, P.J. & McCoy, C.E. (1994). Design of a cooperative problem-solving system for en-route flight planning: An empirical evaluation. *Human Factors, 36,* 94-119.

Lerner, J. S., & Tetlock, P. E. (1999). Accounting for the effects of accountability. *Psychological Bulletin, 125,* 255-275.

Leveson, N. G. & Turner, C. S. (1993). An investigation of the Therac-25 accidents. *Computer*, July, 18-41.

Leveson, N. G. (2001). Systemic factors in software-related spacecraft accidents. American Institute of Aeronautics and Astronautics, AIAA.

Lipshitz, R. (2000). *There is more to seeing than meets the eyeball: The art and science of observation.* In Proceedings of The 5th International Conference on Naturalistic Decision Making, May 26-28, Tammsvik, Sweden.

Lin, L., Isla, R., Doniz, K., Harkness, H., Vicente, K. J. & Doyle, D. J. (1998). Applying human factors to the design of medical equipment: Patient-controlled analgesia. *Journal of Clinical Monitoring, 14*, 253-263.

Lind, M. (2003). Making sense of the abstraction hierarchy in the power plant domain. *Cognition Technology and Work, 5*, 67-81.

Lopes, L. S. et al. (2001). Sentience in robots: Applications and challenges. *IEEE Intelligent Systems, 16*, 66-69.

Malin, J., Schreckenghost, D., Woods, D., Potter, S., Johannesen, L., Holloway, M. & Forbus, K. (1991) *Making intelligent systems team players: Case studies and design issues.* (NASA Tech Memo 104738). Houston, TX: NASA Johnson Space Center.

Mark, G. (2002). Extreme collaboration. *Communications of the ACM. 45*, 89-93.

McCabe, K. (2003). A Cognitive Theory of Reciprocal Exchange. In E. Ostrom & J. Walker (Eds.), *Trust and reciprocity: Interdisciplinary lessons from experimental research.* Russell Sage Foundation, NY.

Mitroff, I. (1974). On systemic problem solving and the error of the third kind. *Behavioral Science, 19*, 383-393.

Moll van Charante, E., Cook, R.I. Woods, D.D. Yue L. & Howie. M.B. (1993). Human-computer interaction in context: Physician interaction with automated intravenous controllers in the heart room. In H.G. Stassen, (Ed.), *Analysis, design and evaluation of man-machine systems 1992*, New York: Pergamon Press.

Moore, C. & Dunham, P., (Eds.) (1995). *Joint attention: its origins and role in development.* Hillsdale, NJ: Lawrence Erlbaum.

Murphy, R. R. (2004). Human-robot interaction in rescue robotics. *IEEE SMC Part C, 34*, 138-153.

Murphy, R. R. & Burke, J. (2005). Up from the rubble: Lessons learned about HRI from search and rescue. In Proceedings of the Human Factors and Ergonomics Society 49th Annual Meeting. 26-28 September, Orlando, FL.

Murray, C. & Cox, C. B. (1989). *Apollo, The race to the moon.* New York: Simon & Schuster.

Murray, J. H. (2005). Narrative abstraction for organizational simulations. In W. Rouse & K. Boff (Ed.). *Organizational simulation*, New York: Wiley.

Myers, K., & Morley, D. (2003). Directing agents. In H. Hexmoor, C. Castelfranchi & R. Falcone (Eds.), *Agent autonomy.* Dordrecht, The Netherlands: Kluwer.

NASA, Mars Climate Orbiter Mishap Investigation Board. (2000). Report on Project Management at NASA, March 13, 2000.

Nass, C., & Moon, Y. (2000). Machines and mindlessness: Social responses to computers. *Journal of Social Issues, 56*, 81-103.

Neisser, U. (1976). *Cognition and reality*. San Francisco: W.H. Freeman.

Neisser, U. (1991). A case of misplaced nostalgia. *American Psychologist, 46*, 34-36.

Nemeth, C., Nunnally, M. O'Connor, M., Klock, P. A. & Cook R. I. (2005). Making information technology a team player in safety: The case of infusion devices. In Henricksen, K., Battles, J., Marks, E. & Lewin, D. (Eds.), *Advances in Patient Safety: From research to implementation. Vol. 1*, Washington, DC: Agency for Healthcare Research and Quality.

Newell, A. (1973). You can't play twenty questions with nature and win. In W.G. Chase (Ed.) *Visual information processing*. New York: Academic Press.

Nikolic, M. I. & Sarter, N. B. (2001). Peripheral visual feedback: A powerful means of supporting attention allocation and human-automation coordination in highly dynamic data-rich environments. *Human Factors, 43*, 30-38.

Nikolic, M. I., Orr, J.. M. & Sarter, N. B. (2004). Why pilots miss the green box: How display context undermines attention capture. *International Journal Of Aviation Psychology, 14*, 39–52.

Nikolic, M. I. & Sarter, N. B. (in press). Disturbance management on modern flight decks: A simulator study of the diagnosis and recovery from breakdowns in pilot-automation coordination. *Human Factors*.

Nisbett, R. & Wilson, T. (1977). Telling more than we know: Verbal reports on mental processes. *Psychological Review 84*, 231-59.

Norman, D. A. (1988). *The psychology of everyday things*. New York: Basic Books.

Norman, D.A. (1990). The 'problem' of automation: Inappropriate feedback and interaction, not 'over-automation.' *Philosophical Transactions of the Royal Society of London, B 327*, 585-593.

Norman, D.A. (1993). *Things that make us smart*. Reading MA.: Addison-Wesley.

Norros, L. (2004). *Acting under uncertainty: The Core-Task Analysis in ecological study of work*. VTT Publication 546. Espoo Finland: Julkaisija Utgivare.

Norros, L. & Nuutinen, M. (2005). Performance-based usability evaluation of a safety information and alarm system. *International Journal of Human-Computer Studies, 63*, 328-361.

Nunnally, M., Nemeth, C. P., Brunetti, V. & Cook, R. I. (2004). Lost in Menuspace: User Interactions with Complex Medical Devices. *IEEE SMC Part A, 34*, 736-742.

Nyssen, A. S. & Javaux, D. (1996). Analysis of synchronization constraints and associated errors in collective work environments, *Ergonomics, 39*, 1249-1264.

Nyssen, A. S. & De Keyser, V. (1998). Improving training in problem solving skills: Analysis of anesthetists' performance in simulated problem situations. *Le Travail Humain, 61*, 387-402.

Olson, G. M., Malone, T. W. & Smith, J. B. (2001). *Coordination theory and collaboration technology*. Hillsdale, NJ: Lawrence Erlbaum.

Ostrom E. (1990). Governing the commons: The evolution of institutions for collective action. New York: Cambridge University Press.

Ostrom, E. (2003). Toward a behavioral theory linking trust, reciprocity, and reputation. In E. Ostrom & J. Walker (Eds.), *Trust and reciprocity: Interdisciplinary lessons from experimental research*. New York: Russell Sage Foundation.

Patterson, E. S., Watts-Perotti, J.C. & Woods, D. D. (1999). Voice loops as coordination aids in Space Shuttle Mission Control. *Computer Supported Cooperative Work: The Journal of Collaborative Computing, 8*, 353-371.

Patterson, E.S., & Woods, D.D. (2001). Shift changes, updates, and the on-call model in space shuttle mission control. *Computer Supported Cooperative Work: The Journal of Collaborative Computing, 10*, 317-346.

Patterson, E.S., Roth, E. M. & Woods, D.D. (2001). Predicting vulnerabilities in computer-supported inferential analysis under data overload. *Cognition, Technology and Work, 3*, 224-237.

Patterson, E.S., Cook, R.I., Render, M.L. (2002). Improving patient safety by identifying side effects from introducing bar coding in medication administration. *Journal of the American Medical Informatics Association, 9*, 540-553.

Patterson, E. S., Cook, R. I., Woods, D.D. & Render, M.L. (2004). Examining the complexity behind a medication error: Generic patterns in communication. *IEEE SMC Part A, 34*, 749-756.

Patterson, R. D. (1990). Auditory warning sounds in the work environment. *Philosophical Transactions of the Royal Society of London, B 327*, 485-492.

Peirce, C. S. (1955). Abduction and induction. In J. Buchler (Ed.), *Philosophical writings of Peirce*. London: Dover, p. 150-156. (Original work published, 1903).

Perkins, D.N. (1992). The topography of invention. In R. J. Weber & D. N. Perkins (Eds.), *Inventive minds*, New York: Oxford University Press.

Potter, S. S., Ball, R. & Elm, W. (1996). Coping with the complexity of aeromedical evacuation planning: Implications for the development of decision support systems. In *Proceedings of the '96 Symposium on Human Interaction with Complex Systems*. Dayton, OH: IEEE.

Potter, S. S., Roth, E. M., Woods, D. D. & Elm, W. (2000). Bootstrapping multiple converging cognitive task analysis techniques for system design. In J.M.C. Schraagen, S.F. Chipman, & V. L. Shalin (Eds.), *Cognitive task analysis*. Hillsdale, NJ: Lawrence Erlbaum.

Potter, S. S., Gualtieri, J. & Elm, W. (2003). Case studies: Applied cognitive work analysis in the design of innovative decision support. In E. Hollnagel (Ed.), *Handbook of cognitive task design*. Hillsdale, NJ: Lawrence Erlbaum.

Rae, A., Jackson, D., Ramanan, P., Flanz, J. & Leyman, D. (2003). Critical feature analysis of a radiotherapy machine. Proceedings of 22nd International Conference on Computer Safety, Reliability and Security (SafeComp) Edinburgh, Scotland, September 2003.

Rasmussen, J. (1979). On the structure of knowledge: A morphology of mental models in a man-machine system context. Risø-M-2192. Electronics Department, Risø National Laboratory, Roskilde, Denmark.

Rasmussen, J. & Lind M. (1981). Coping with complexity. In H. G. Stassen (Ed.), *First European annual conference on human decision making and manual control*. New York: Plenum. (Also as Risø-M-2293. Electronics Department, Risø National Laboratory, Roskilde, Denmark 1981).

Rasmussen, J. & Rouse, W. B. (1981). *Human detection and diagnosis of system failures*. New York: Plenum.

Rasmussen, J. (1985). Trends in human reliability analysis. *Ergonomics, 28*, 1185-1196.

Rasmussen, J. (1986). Information processing and human-machine interaction: An approach to Cognitive Engineering. New York: North-Holland.

Rasmussen, J., Pejtersen, A. M., & Goodman, L. P. (1994). *Cognitive Systems Engineering*. New York: Wiley.

Reason, J. (1990). *Human error*. Cambridge, England: Cambridge University Press.

Reeves, B. & Nass, C. (1996). The media equation: How people treat computers, televisions, and new media like real people and places. New York : Cambridge University Press.

Reiersen, C. S., Marshall, E. & Baker, S. M. (1988). An experimental evaluation of an advanced alarm system for nuclear power plants. In J. Patrick & K. Duncan (Eds.), *Training, human decision making and control*. New York: North-Holland.

Remington, R. W. & Shafto, M. G. (1990). Building human interfaces to fault diagnostic expert systems I: Designing the human interface to support cooperative fault diagnosis. In *CHI '90 Workshop on Computer-Human Interaction in Aerospace Systems*. Seattle, WA., April 1-2.

Rochlin, G. I. (1999). Safe operation as a social construct. *Ergonomics, 42*, 1549-1560.

Roesler, A., Feil, M. & Woods, D.D. (2001). Design is telling (sharing) stories about the future. MediaPaper, Cognitive Systems Engineering Laboratory, The Ohio State University. url: http://csel.eng.ohio-state.edu/animock.

Roesler, A., Woods, D. D. & Feil, M. (2005). Inventing the future of cognitive work. In W. Jonas, R. Chow, N. Verhaag (Eds.), Design-System-Evolution: Application of systemic and evolutionary approaches to design theory, design practice, design research and design education. Bremen, Germany: European Academy of Design.

Roth, E.M., Bennett, K. & Woods. D.D. (1987). Human interaction with an 'intelligent' machine. *International Journal of Man-Machine Studies, 27*, 479-525.

Roth, E.M., Woods, D.D. & Pople, H.E. (1992). Cognitive simulation as a tool for cognitive task analysis. *Ergonomics, 35*, 1163-1198.

Roth, E. M., Malin, J. T. & Schreckenghost, D. L. (1997). Paradigms for intelligent interface design. In M. Helander, T. Landauer & P. Prabhu (Eds.) *Handbook of human-computer interaction* (2nd Ed), Amsterdam: North-Holland.

Rudolph, J. (2003). Into the Big Muddy and Out Again: Error persistence and crisis management in the operating room. Unpublished Doctoral dissertation. Boston College, Chestnut Hill, MA.

Salomon, G. (1991). Transcending the qualitative-quantitative debate: The analytic and systemic approaches to educational research. *Educational Researcher*, August-September, 10-18.

Sarter, N. B. & Woods, D. D. (1995). How in the world did we get into that mode? Mode error and awareness in supervisory control. *Human Factors*, *37*, 5-19.

Sarter, N. B. & Woods, D. D. (1997). Team play with a powerful and independent agent: A corpus of operational experiences and automation surprises on the Airbus A-320. *Human Factors*, *39*, 553-569.

Sarter, N. B., Woods, D. D. & C. Billings, C. (1997). Automation surprises. In G. Salvendy, (Ed.), *Handbook of human factors/ergonomics* (2nd Edition), New York: Jon Wiley & Sons.

Sarter, N.B. & Woods, D.D. (2000). Team play with a powerful and independent agent: A full mission simulation. *Human Factors*, *42*, 390-402.

Sarter, N. B. & Amalberti, R., editors. (2000). *Cognitive engineering in the aviation domain*, Hillsdale, NJ: Lawrence Erlbaum.

Sarter, N. B. (2002). Multimodal information presentation in support of human-automation communication and coordination. In E. Salas (Ed.), *Advances in human performance and cognitive engineering research*. New York: JAI Press.

Schoenwald, J., Trent, S., Tittle, J. & Woods, D. D. (2005). Scenarios As A Tool For Collaborative Envisioning: Using The Case of New Sensor Technologies for Military Urban Operations. Proceedings of the Human Factors and Ergonomics Society 49th Annual Meeting. 26-28 September, Orlando FL. [see url: http://csel.eng.ohio-state.edu/productions/xcta]

Scholl, B. J. & Tremoulet, P. D. (2000). Perceptual causality and animacy. *Trends in Cognitive Science*, *4*, 299-309.

Schön, D. A., (1983). *The reflective practitioner*. New York: Basic Books.

Shalin, V. L. (2005). The roles of humans and computers in distributed planning for dynamic domains. *Cognition Technology and Work*, *7*, 198-211.

Sharpe, V. A. (2004). Accountability and justice in patient safety reform. In V. A. Sharpe (ed.), *Accountability: Patient safety and policy reform*. Washington DC: Georgetown University Press.

Shattuck, L. G., & Woods, D. D. (2000). Communication of intent in military command and control systems. In C. McCann & R. Pigeau (Eds.), *The human in command: Exploring the modern military experience*. New York: Plenum.

Sheridan, T. B. (1992). Telerobotics, automation, and human supervisory control. Cambridge, MA: M.I.T. Press.

Simon, H. A. (1969). *The sciences of the artificial*. Cambridge, MA: M.I.T. Press.

Sklar, A.E. & Sarter, N.B. (1999). "Good Vibrations": The use of tactile feedback in support of mode awareness on advanced technology aircraft. *Human Factors*, *41*, 543-552.

Smith, P. J., McCoy, E. & Layton, C. (1997). Brittleness in the design of cooperative problem-solving systems: The effects on user performance. *IEEE Transactions on Systems, Man and Cybernetics*, *27*, 360-371.

Smith, P.J., Woods, D., McCoy, E., Billings, C., Sarter, N., Denning, R. & Dekker, S. (1998). Using forecasts of future incidents to evaluate future ATM system designs. *Air Traffic Control Quarterly*, *6*, 71-86.

Smith, P. J., McCoy, E. & Orasanu, J. (2001). Distributed cooperative problem-solving in the air traffic management system. In G. Klein & E. Salas (Eds.), *Naturalistic decision making.* Hillsdale, NJ: Erlbaum.

Smith, P. J., Beatty, R., Spencer, A. & Billings, C. (2003). Dealing with the challenges of distributed planning in a stochastic environment: Coordinated contingency planning. Proceedings of the 2003 Annual Conference on Digital Avionics Systems, Chicago, IL.

Smith, P. J., Klopfenstein, M., Jezerinac, J. & Spenser, A. (2004). Distributed work in the National Airspace System: Providing feedback loops using the post-operations evaluation tool (POET). In B. Kirwan, M. Rodgers & D. Schaefer (Eds.), *Human factors impacts in air traffic management.* London: Ashgate.

Sorkin, R. D. & Woods, D. D. (1985). Systems with human monitors: A signal detection analysis. *Human-Computer Interaction, 1,* 49-75.

Suchman, L. A. (1987). *Plans and situated actions: The problem of human-machine communication.* Cambridge, England: Cambridge University Press.

Theureau, J. (2003). Course-of-action analysis and course-of-action centered design. In E. Hollnagel (Ed.), *Handbook of cognitive task design.* Hillsdale, NJ: Lawrence Erlbaum.

Teigen, K. H. & Keren, G. (2003). Surprises: low probabilities or high contrasts? *Cognition, 87,* 55-71.

Tetlock, P. E. (1999). Accountability theory: Mixing properties of human agents with properties of social systems. In L. L. Thompson, J. M. Levine, & D. M., Messick, (Eds.), *Shared cognition in organizations: The management of knowledge.* Hillsdale, NJ: Lawrence Erlbaum.

Toulmin, S., Rieke, R. & Janik, A. (1984). *An introduction to reasoning.* New York: Macmillan.

Tukey, J. W. (1977). *Exploring data analysis.* Reading, MA: Addison-Wesley.

U.S. Air Force Aircraft Accident Investigation Board Report (1999). RQ4A Global Hawk UAV, 98-2003, Edwards AFB, CA.

Vicente, K. (1999). Cognitive Work Analysis: Toward safe productive and healthy computer based work. Hillsdale, NJ: Lawrence Erlbaum.

Von Uexkull, J. (1934). A stroll through the worlds of animals and men. Reprinted in C. Schiller (Ed.), *Instinctive behavior.* New York: International Universities Press, 1957.

Warren. W. H. & Shaw, R. E. (1985). Events and encounters as units of analysis for ecological psychology. In W. H Warren & R. E. Shaw (Eds.), *Persistence and change.* Hillsdale, NJ: Lawrence Erlbaum.

Watts, J., Woods, D. D. & Patterson, E. S. (1996). Functionally distributed coordination during anomaly response in space shuttle mission control. Proceedings of Human Interaction with Complex Systems, IEEE Computer Society Press, Los Alamitos, CA.

Watts-Perotti, J. &. Woods, D. D. (1999). How experienced users avoid getting lost in large display networks. *International Journal of Human-Computer Interaction, 11,* 269-299.

Wears, R. L. & Berg, M. (2005). Computer technology and clinical work: Still waiting for Godot. *JAMA, 293(10),* 1261-1263.

Weick, K. E., Sutcliffe, K. M. & Obstfeld, D. (1999). Organizing for High Reliability: Processes of Collective Mindfulness. *Research in Organizational Behavior, 21*, 81-123.

Wiener, N. (1948). Cybernetics, or, control and communication in the animal and the machine. New York: Wiley.

Wiener, N. (1950). The human use of human beings: Cybernetics and society. New York: Doubleday.

Wiener, N. (1964). *God and golem.* Cambridge, MA: M.I.T. Press.

Wiener, E. L. (1989). *Human factors of advanced technology ("glass cockpit") transport aircraft.* (NASA Contractor Report No. 177528). Moffett Field, CA: NASA-Ames Research Center.

Winograd, T. & Flores, F. (1986). *Understanding computers and cognition.* Norwood, NJ, Ablex.

Woods, D.D. (1984). Visual Momentum: A concept to improve the cognitive coupling of person and computer. *International Journal of Man-Machine Studies, 21*, 229-244.

Woods, D.D. (1986a). Cognitive Technologies: The design of joint human-machine cognitive systems. *AI Magazine, 6*, 86-92.

Woods, D.D. (1986b). Paradigms for intelligent decision support. In E. Hollnagel, G. Mancini, & D.D. Woods (Eds.), *Intelligent decision support in process environments.* New York: Springer-Verlag.

Woods, D. D. & Hollnagel, E. (1987). Mapping cognitive demands in complex problem solving worlds. *International Journal of Man-Machine Studies, 26*, 257-275.

Woods, D. D., O'Brien, J. & Hanes, L. F. (1987). Human factors challenges in process control: The case of nuclear power plants. In G. Salvendy, editor, *Handbook of human factors/ergonomics* (1st Ed), New York: Wiley.

Woods, D. D. (1988). Coping with complexity: The psychology of human behavior in complex systems. In L.P. Goodstein, H.B. Andersen & S.E. Olsen (Eds.), *Mental models, tasks and errors*, London: Taylor & Francis.

Woods, D. D. & Roth, E.M. (1988). Cognitive Systems Engineering. In M. Helander (Ed.), *Handbook of human-computer interaction* (1st Ed), North-Holland, New York. (Reprinted in N. Moray, editor, *Ergonomics: Major writings.* Taylor & Francis, 2004.)

Woods, D. D., Roth, E.M. & Bennett, K.B. (1990). Explorations in joint human-machine cognitive systems. In S. Robertson, W. Zachary, & J. Black, (Eds.), *Cognition, computing and cooperation.* Norwood, NJ: Ablex.

Woods, D. D. (1993). Process-tracing methods for the study of cognition outside of the experimental psychology laboratory. In G. Klein, J. Orasanu, R. Calderwood, & C. E. Zsambok, C. E. (Eds.), *Decision making in action: Models and methods.* Norwood, NJ: Ablex.

Woods, D.D. & Sarter, N. (1993). Evaluating the Impact of New Technology on Human-Machine Cooperation. In J. Wise, V. D. Hopkin, & P. Stager, (Eds.), *Verification and Validation of Complex Systems: Human Factors Issues.* Berlin: Springer-Verlag.

Woods, D. D. (1994). Cognitive demands and activities in dynamic fault management: Abduction and disturbance management. In N. Stanton, (Ed.), *Human factors of alarm design*, London: Taylor & Francis.

Woods, D. D., L. Johannesen, R.I. Cook & N. Sarter. *Behind human error: Cognitive systems, computers and hindsight.* (1994). Dayton OH: Crew Systems Ergonomic Information and Analysis Center, WPAFB.
(order at http://iac.dtic.mil/hsiac/SOARS.htm#Past)

Woods, D. D. (1995a). Towards a theoretical base for representation design in the computer medium: Ecological perception and aiding human cognition. In J. Flach, P. Hancock, J. Caird, & K. Vicente, (Eds.), *An ecological approach to human machine systems I: A global perspective.* Hillsdale, NJ: Lawrence Erlbaum.

Woods, D. D. (1995b). The alarm problem and directed attention in dynamic fault management. *Ergonomics, 38*, 2371-2393.

Woods, D. D. (1996). Decomposing automation: Apparent simplicity, real complexity, In R. Parasuraman & M. Mouloua, (Eds.), *Automation technology and human performance*, Hillsdale, NJ: Lawrence Erlbaum.

Woods, D. D. & Watts, J. C. (1997). How not to have to navigate through too many displays. In Helander, M.G., Landauer, T.K. & Prabhu, P. (Eds.), *Handbook of human-computer interaction* (2nd Ed), Amsterdam: Elsevier Science.

Woods, D. D. (1998). Designs are hypotheses about how artifacts shape cognition and collaboration. *Ergonomics, 41*, 168-173.

Woods, D.D. & Tinapple, D. (1999). *W3: Watching Human Factors Watch People at Work.* Presidential Address, 43rd Annual Meeting of the Human Factors and Ergonomics Society, September 28, 1999. Multimedia Production at url http://csel.eng.ohio-state.edu/hf99/

Woods, D.D. & Dekker, S. W. A. (2000). Anticipating the effects of technological change: a new era of dynamics for Human Factors. *Theoretical Issues in Ergonomic Science, 1*, 272-282.

Woods D. D. & Patterson, E. S. (2000). How unexpected events produce an escalation of cognitive and coordinative demands. In P. A. Hancock & P. Desmond (Eds.), *Stress, workload and fatigue.* Hillsdale, NJ: Lawrence Erlbaum.

Woods, D.D. & Sarter, N. (2000). Learning from automation surprises and going sour accidents. In N. Sarter & R. Amalberti (Eds) *Cognitive Engineering in the aviation domain.* Hillsdale, NJ: Lawrence Erlbaum.

Woods D. D. & Shattuck L. G. (2000). Distant supervision—local action given the potential for surprise. *Cognition, Technology and Work 2*, 86-96.

Woods, D. D. & Christoffersen, K. (2002). Balancing practice-centered research and design. In M. McNeese & M. A. Vidulich (Eds.), *Cognitive systems engineering in military aviation domains.* Wright-Patterson AFB, OH: Human Systems Information Analysis Center.

Woods, D.D. (2002). Steering the reverberations of technology change on fields of practice: Laws that govern cognitive work. In Proceedings of the 24th Annual

Meeting of the Cognitive Science Society, Atlanta, GA. [see url: http://csel.eng.ohio-state.edu/laws]

Woods, D.D. & Cook, R.I. (2002). Nine steps to move forward from error. *Cognition, Technology, and Work, 4*, 137-144.

Woods, D. D., Patterson, E. S., & Roth, E. M. (2002). Can we ever escape from data overload? A cognitive systems diagnosis. *Cognition, Technology, and Work, 4*, 22-36.

Woods, D. D. (2003). Discovering How Distributed Cognitive Systems Work. In E. Hollnagel (Ed.), *Handbook of cognitive task design*. Hillsdale, NJ: Lawrence Erlbaum.

Woods, D. D., Tittle, J., Feil, M. & Roesler, A. (2004). Envisioning human-robot coordination for future operations. *IEEE SMC Part C, 34*, 210-218.

Woods D. D. (2005a). Conflicts between learning and accountability in patient safety. *DePaul Law Review, 54*, 485-502.

Woods, D. D. (2005b). Creating foresight: Lessons for resilience from Columbia. In W. Starbuck & M. Farjoun (Eds.), *Organization at the Limit: NASA and the Columbia Disaster*. Malden, MA: Blackwell.

Woods, D. D. (2005c). Supporting cognitive work: How to achieve high levels of coordination and resilience in joint cognitive systems. Naturalistic Decision Making 7. Amsterdam, The Netherlands, June 15, 2005.

Woods, D. D. (2006a). Essential characteristics of resilience for organizations. In E. Hollnagel, D. D. Woods & N. Leveson, (Eds.), *Resilience Engineering: Concepts and precepts*. Aldershot, U.K.: Ashgate.

Woods, D. D. (2006b). How to design a safety organization: Test case for resilience engineering. In E. Hollnagel, D.D. Woods & N. Leveson, (Eds.), *Resilience Engineering: Concepts and precepts*. Aldershot, U.K.: Ashgate.

Woods, D. D. & Cook, R. I. (2006). Incidents: Are they markers of resilience or brittleness? In E. Hollnagel, D. D. Woods & N. Leveson, (Eds.), *Resilience Engineering: Concepts and precepts*. Aldershot, U.K.: Ashgate.

Xiao, Y., Seagull, J., Nieves-Khouw, F., Barczak, N. & Perkins, S. (2004). Organizational-historical analysis of the "failure to respond to alarm" problems. *IEEE SMC Part A, 34*, 772-778.

Zacks, J. M. (2004). Using movement and intentions to understand simple events. *Cognitive Science, 28*, 979-1008.

Zhang, J. & Norman, D. A. (1994). Representations in distributed cognitive tasks. *Cognitive Science, 18*, 87-122.

Appendix A

Verbal Protocol for the Bradycardia Update Case

(FROM JOHANNESEN ET AL., 1994)

A = attending Anesthesiologist; R = Senior Resident.

TRANSCRIPT	DOMAIN-INDEPENDENT DESCRIPTION	PROBLEM-SOLVING PROCESS
A: *{enters room}* Nice and tachycardic[1]	A comments on process	
R: Yeah, well, better than nice and bradycardic …		
A: What's going on guys?	A makes open-ended request for update	
R: *{takes end of printout, seems to show to A}* She had an episode of just kinda, all of the sudden bradying down to 50, 52 then came right back up, nothing they were doing, then all of the sudden out of the blue, I was shooting an output[2] and she dropped down to 32, 38[3] somewhere around there, pressure[4] dropped down to 60 so I gave her .5 of atropine[5] and ah, kicked her up to 6.5; she liked that, but no explanation. This is at 50 millimeters per second, twice the speed.[6]	R mentions: -previous related event, including dynamics and approx values -discounting of other agents' activities as cause -action taken while event occurred - dynamics and approx values of relevant parameter during event -corrective action taken and process' response -no good candidate for diagnostic search R supplements description with artifact preserving data history	Initial update of significant event

TRANSCRIPT	DOMAIN-INDEPENDENT DESCRIPTION	PROBLEM-SOLVING PROCESS
A: They weren't in the head doing anything?	A requests specific past observation information (concerning other agents' activities) at time of event.	Hypothesis building
R: Nothing.	R answer discounts hypothesis, but does not elaborate.	
A: Okay. Well I can't necessarily		
R: The only thing		
A: I can't necessarily explain that	A states has no candidates	
R: Yeah, neither can I. The only thing we're doing right now is just trying to open her up and fill her up. {points to right IV tree} She's up to a mic per kilo of nitro[7] and then she's still at the 5, started out at 3 and a half of dobutamine[8] and it did absolutely nothing, so I'm up to 5	R provides more information on current actions and previous actions	Context building
A: Okay		
R: So I don't know if she doesn't like contractility or, I can't think of anything else we're doing. The line went in perfectly normal, I can't imagine that she has a pneumo or anything that would be causing tension, her peak area pressures have not changed. Just all of the sudden -boom-out of the blue--her potassium is 3 point 3 and we're getting ready to replace that and we have been hyper-ventilating, but I don't know if low potassium can affect heart rate	R offers hypothesis but discounts based on his knowledge R offers another hypothesis but discounts it based on data Dynamics of event repeated Process variables mentioned, action to be taken mentioned R offers a third hypothesis but voices his lack of knowledge	Hypothesis discounting

TRANSCRIPT	DOMAIN-INDEPENDENT DESCRIPTION	PROBLEM-SOLVING PROCESS
A: Yeah, I don't know, I can't give you cause and effect on that. In my experience it's usually been stimulation of the trachea, it's something traction on the dura	A mentions two causes of the significant event based on his past experience	Case-based discussion
R: Yeah, (absolutely)		
A: You know things		
R: Yeah, it may have been dura	R remarks that one of these causes may have been cause in this case	
A: ...Sort of a reflex, pressure on an eye	A provides another possible cause based on past cases	
R: {animated} Actually it was when they were sawing the dura open.	R remarks that event occurred during a time when one of the causes mentioned by A could have occurred	Discounted hypothesis reconsidered
A: Well that's ... R: Putting tension on it		
R: Traction on the dura		
A: You touch the dura you'll get that	A states mechanism	
R: Okay		
A: Cause the dura is ennervated by the fifth I believe, and it somehow makes its way back to the (.) ganglion, same thing that causes oculocardiac reflex.	A describes mechanism whereby hypothesized cause leads to the significant event	
R: I'd be willing to bet you're absolutely right {R waves pen over ventilator setting knobs, then leaves view}	R expresses confidence for hypothesis	Hypothesis acceptance

TRANSCRIPT	DOMAIN-INDEPENDENT DESCRIPTION	PROBLEM-SOLVING PROCESS
A: Is the same mechanism whereby you get (bradycardial traction) on the dura, so my guess is that's exactly what it was	A continues explanation of mechanism	
R: Okay.	R concurs (with hypothesis)	
A: You now and for future reference, if you suspect {pause} this lady's probably not going to mind this experience because she, we don't think she's really significantly sick, we're being a little overly cautious with her, my preference is, if you have a patient that you think has a bad heart, and you think they have a vagal problem via traction, or an eye...		
R: So that's why		
A: It's traction on the dura		

[1] Tachycardia refers to rapid heart rate, while bradycardia refers to a slow heart rate.

[2] Cardiac output refers to the volume of blood per unit time that the heart moves. The measurement of cardiac output requires injection of a measured amount of IV fluid and is done infrequently.

[3] These are very low heart rate values, requiring treatment.

[4] blood pressure.

[5] A drug that increases heart rate by blocking the parasympathetic system.

[6] Chart speed for EKG recording is usually 25 mm/sec. Because it's running at 50mm/sec, recorded events occupy twice the length of chart paper than they would at normal speed.

[7] Nitroglycerine. A vasodilator, for controlled hypotension.

[8] Dobutamine is generally given for low cardiac output, in order to increase contractility.

Appendix B
Adapting to New Technology

New technology transforms roles, judgments, vulnerabilities, relationships as people adapt to exploit capabilities and workaround complexities. The new artifacts represent new forms of activity with new requirements for support to meet the new or changed demands on practice. Here is an example from NASA mission control following a new round of computerization that provided new display capabilities and some new automation for monitoring systems on the space shuttle. The case was captured compactly for us, when a reflective practitioner (the section manager who was also an experienced practitioner) shared his examination of the impact of the new systems on operations with his staff.

> **Practitioner reflections on impact of new technology in mission control.**
> Now that we have three flights from the CCC [mission control], I wanted to comment on some things regarding console operations from the new workstation consoles.
>
> First, I have been very impressed by the performance of the CCC and the applications. I thought the transition went very smoothly. ... Now that we have all this neat stuff what does it mean? We have much more flexibility in how our displays look and in the layout of the displays on the screens. We also have the added capabilities that allow the automation of the monitoring of telemetry.
>
> A general rule of life is that when something has advantages, it usually has disadvantages, and the new consoles are no exception. This fact was made evident to me by my observations of operations during flights and sims, both as an operator and an evaluator. Other group members have also mentioned concerns to me as well. What I think it boils down to is two potential gotchya's, these are:
>
> 1. Too much flexibility. There is so much stuff to play with that it can get to the point where adjusting stuff on the console distracts from keeping up with operations. The configuration of the displays, the various supporting applications, the ability to "channel surf" on the TV, all lead to a lack of attention to operations. I have seen teams miss events or not hear calls on the loops because of being preoccupied with the console. I have also witnessed that when a particular application doesn't work, operations were missed due to trying to troubleshoot the problem. And this was an application that was not critical to the operations in progress. ...
>
> 2. Too much reliance on automation, mainly, the Telemetry Monitor program. When we started Telemetry Monitor it was intended to catch changes during the long periods of no RMS [remote manipulator system—the robot arm] operations and to assist in monitoring events as they occur. I'm concerned that it is becoming

209

the prime (and sometimes sole) method for following operations. During operations, the displays, telemetry monitor, voice loops and TV downlink when available should be used to track operations. All these sources provide information in a unique way and they back each other up. When the crew is taught to fly the arm, they are trained to use all sources of feedback, the D&C panel, the window views, multiple camera views, and the spec. When we "fly" the console, we must do the same. This point was made very evident during a recent sim[ulation] when Telemetry Monitor wasn't functioning. It took the team awhile to notice that it wasn't working because they weren't cross checking and then once they realized it they had some difficulty monitoring operations. If this were to happen in flight it could, at a minimum, be embarrassing to be caught unaware of what was happening, and, at a maximum, lead to an incorrect failure diagnosis or missing a failure or worse—such as a loss of a payload.

The solution to this problem is a simple one. We need to exercise judgment to prioritize tending to the console vs. following operations. If something is required right now, fix it or work around it. If it's not required and other things are going on, let it wait. Secondly, we all need to develop our scan patterns so that all data sources are utilized. The crew is trained to do this when they fly, pilots are trained to do this, and we as flight controllers must recognize the importance of developing a scan pattern and practice it so that it becomes automatic.

    Electronic memo to section, Section Head for Remote Manipulator System, NASA JSC, February 1996.

The Section Head's comments concisely capture some of the core processes in JCSs at work: new demands on control of attention and anticipation, changed mix of workload over time, new demands for judgment, new vulnerabilities for failures to revise assessments (see also Sarter et al., 1997 for similar data from aviation). It is striking to note how the new artifacts and associated capabilities resulted in new demands for expert judgment and focused attention on new aspects of how situations evolve. Implicit in the Section Head's comments is a concern about how to develop and sustain the new demands for expert judgment, since at this time cost pressure was reducing investments in developing experience (e.g., less opportunities for practice). Also note that the memo is in itself data about the adaptation process as people in various roles reflect and modify operations to cope with the complexities that result from change under pressure. For example, the comments reflect how people in various roles are sensitive to new paths toward failure and how they attempt to re-adjust practice to forestall these vulnerabilities (how their strategies are failure sensitive; Woods and Cook, 2002; Hollnagel et al., 2006)—how to sustain resilience.

Author Index

211

Subject
Index

Printed in the United States
by Baker & Taylor Publisher Services